ESSENTIALS OF
HUMAN ANATOMY
AND PHYSIOLOGY

ESSENTIALS OF

HUMAN ANATOMY AND PHYSIOLOGY

SECOND EDITION

BY JAMES PALMER
SOUTHEASTERN LOUISIANA UNIVERSITY

cognella®
SAN DIEGO

Bassim Hamadeh, CEO and Publisher
Angela Schultz, Senior Field Acquisitions Editor
Carrie Montoya, Manager, Revisions and Author Care
Kaela Martin, Project Editor
Casey Hands, Associate Production Editor
Emely Villavicencio, Senior Graphic Designer
Alexa Lucido, Licensing Supervisor
Natalie Piccotti, Director of Marketing
Kassie Graves, Vice President of Editorial
Jamie Giganti, Director of Academic Publishing

Cover image: Copyright © 2016 iStockphoto LP/yodiyim.
Interior images: Copyright © 2016 iStockphoto LP/yodiyim.
    Copyright © Depositphotos/popcic.
    Copyright © Depositphotos/leshkasmok.
    Copyright © Depositphotos/bioraven.
    Copyright © Depositphotos/alexandragl.
    Copyright © Depositphotos/jehsomwang.
    Copyright © Depositphotos/eveleen.
    Copyright © Depositphotos/Mr. Webicon.

Printed in the United States of America.

cognella® | ACADEMIC PUBLISHING

3970 Sorrento Valley Blvd., Ste. 500, San Diego, CA 92121

BRIEF
CONTENTS

Note to readers: Content in chapters 1 through 7 applies to the content addressed in the first part of a two-session Human Anatomy and Physiology course.

TABLE OF
CONTENTS

Note to readers: Content in chapters 1 through 7 applies to the content addressed in the first part of a two-session Human Anatomy and Physiology course.

GENERAL INTRODUCTION

Human anatomy and physiology is very much like learning a different language. The vocabulary is composed of many terms derived from Greek or Latin. Learning these terms and definitions will make the course easy. Good luck in your efforts and future. These notes consist of the essential knowledge needed for individuals entering paramedical fields and those reviewing basic knowledge for continuing education.

CHAPTER ONE

INTRODUCTION TO HUMAN ANATOMY AND PHYSIOLOGY

LEARNING OBJECTIVES

Upon completion of the chapter readers will have the general knowledge of the concepts of living organisms:

1. The organization of living organisms from the elemental to the complete organism
2. The characteristics living organisms exhibit supporting life processes
3. The processes used to maintain homeostasis
4. The directional terms used to express location and relationship of structures, anatomical position, planes, and divisions of the axial body

General Terms

The first terms you must understand are anatomy and physiology. *Anatomy* is the study of the structure and shape of the body and its parts, usually with the emphasis on how structure influences function. *Physiology* is the study of the functions of the body and the role of various structures on the function. Another term that is very important to understand is *homeostasis*, which is the balance or equilibrium of various functions of the body, such as temperature, pH, blood pressure, and nutrition, as it maintains the internal environment.

Levels of Organization of Living Organisms

A living organisms' survival depends on maintaining an organized series of structures. It begins at the *chemical level* of organization, where various atoms or elements are organized into molecules, forming structural and functional components required to support the next higher level. The *cellular level* of organization is where the molecules are used to provide the structure and function required for life. Groups of cells having similar structure and function are organized into the *tissue level* of organization, where these tissues perform specific functions required to support life. Groups of tissues organized into structure with a defined shape and function make up the *organ level* of organization, with examples such as the liver, heart, lungs, stomach, and others. Groups of tissue coupled with one or more organs are organized into a *systems level*, which performs a series of various functions supporting the organism; for example, the urinary, cardiovascular, and the nervous systems. The final level of organization to consider is the *organism level*, which is all of the systems working together to make the living organism function in the environment.

Life Processes

In order for life to exist, a series of events and processes must occur. The ability of an organism to control energy is called *metabolism*, which is the sum of all chemical activities within the body. It is divided into those activities which harvest energy, catabolic processes, and those which transform and store energy, anabolic processes. The organism requires the ability to detect changes in its internal and external environments, and respond to those changes; this is *responsiveness*. If the organism's materials must be relocated to meet needs both in the organism and within the cells, this is *movement*, which is vital for the removal of wastes and nutrients and the removal of the organism from threats. A living organism's increase in size or number of cells defines *growth*, which is critical for the development of the organism through life. Since organisms develop from a single unspecialized cell, which provides the basis of all the specialized cells, the process of *differentiation* occurs, where the single zygote gives rise to muscle, bone, and blood. The final life process that is essential for the survival of organisms is *reproduction*, which is required for repair and replacement of damaged cells and tissues as well as the production of a new individual.

Homeostasis

Homeostasis is the outcome of the processes that maintain internal balance or equilibrium in the body. It is controlled by the actions of specific control centers in the brain and the actions of hormones produced by the endocrine system. Changes are made in various systems in response to stresses placed on the organism, either external or internal. When the balance is disrupted, the body will use mechanisms to attempt to reduce the effect of the stress, which is *negative feedback*, or enhance the response of the body to the stress, which is positive feedback. An example of negative feedback would be as follows: after a meal your blood sugar increases as you digest the food; in response, your pancreas releases insulin, which promotes the cells to absorb the glucose and the liver and muscle cells to convert it to glycogen. An example of *positive feedback* would be as follows: a person who is allergic is exposed to a sting, and the person's immune response magnifies the level of histamines released, causing a risk of death. Another positive feedback example would be a woman in labor, where the endocrine system releases oxytocin, which increases the strength of the contraction of muscles to aid in expelling the fetus.

Directional Terms

When we discuss various structures on the body or the relationship of structures, establishing a reference position is necessary. This is called the *anatomical position*, which is defined as either standing erect, facing forward with the palms of the hands forward, or supine (laying on the back facing upward) with the palms facing upward. This puts various body structures in a predictable relationship. We frequently use a variety of terms to describe how structures relate. If a structure is above a defined reference point, we would describe it as *superior* to that structure. If it is below the reference point, it would be *inferior*. Structures defined as being toward the front of the body would be *anterior* or *ventral*; toward the back would be *posterior* or *dorsal*. The midline of the body or structure along the long being the reference point a structure or feature would be either closer to the midline would be *medial* and further away from the midline would be *lateral*. The term *intermediate* may occasionally be used when discussing three structures. When working with the appendages, a structure closer to the central body would be *proximal* and a structure further away would be *distal*. A structure located on or near the surface would be *superficial*, and one closer to the skeleton would be *deep*.

Planes and Cavities

Planes allow the division of the body into defined areas for ease of placement of various structures. The *sagittal plane* divides the body along the long axis into right and left, with various structures located either right or left of the midline. The *frontal (coronal) plane* divides the body into an anterior or posterior relationship. The *horizontal (transverse) plane* divides the structures into superior and inferior relationships.

Cavities of the body were originally used with four-legged animals, but the human rotated into an upright position rearranges those cavities. The *dorsal cavity*, also referred to as the posterior cavity, is composed of various body structures protecting vital structures. It is subdivided into a *cranial component*, which surrounds and encapsulates the brain in bone, and the *spinal component*, which surrounds the spinal cord in bone and muscle. The *ventral cavity* is located in the central body area and is composed of the *thoracic portion*, composed of bone and muscle, separated from the *abdominopelvic portion* by the diaphragm. The thoracic portion is divided into three cavities: two housing the lungs, which are the *pleural cavities*, and the mediastinal region, which contains the *pericardial cavity*, housing the heart. The abdominopelvic portion is divided into an abdominal region, located above the top of the pelvis, and the pelvis proper.

The abdominal area is divided into areas for the purpose of locating various structures, or to suggest which structures might be involved in a patient's complaint. In medicine, the quadrant system is used to divide the abdomen into right and left upper quadrants, and right and left lower quadrants. Anatomists divide the abdomen into nine regions, with the central region being the umbilical (navel) and on either side being the right or left lumbar; above the umbilical in the center would be the epigastric and either the right or left hypochondriac; below the umbilical in the center would be the hypogastric (pubic) and left or right iliac (inguinal) regions.

Systems

The human body can easily be divided into systems, and most agree that there are eleven or twelve systems. The typical coverage of the systems begins with the *integument system* (skin), which covers the exterior of the body, providing protection and sensory input, and acting as a blood reservoir and a site of the synthesis of vitamin D_3, a precursor of vitamin D. The next system is typically the *skeleton system*, composed of bone, cartilage, and other connective tissues providing protection, support, movement, mineral storage, and energy storage. The *muscle system* is described by most texts with a focus on the skeletal muscle, describing the anatomy, organization, and physiology in detail, with limited mention of cardiac and smooth muscle tissue. The *nervous system* is usually the next system covered, with the anatomy and physiology of both the neurons and neuroglia in the central and peripheral systems, coupled with the primary sensory structures. These are normally covered in the first semester of a two-semester course.

The second semester begins with the *circulatory* or *cardiovascular systems*, which cover the blood, the heart, and the vascular systems. The next system covered is usually the *respiratory system*, due to its interactions with the blood and heart to supply oxygen and remove waste carbon dioxide to maintain the pH of the body. The *digestive system* frequently follows, with description of the anatomy and physiology of the components of the system and nutrition. This is followed by the *urinary system*, with the anatomy and physiology contributions to the fluid and electrolyte values in the body. The anatomy and physiology of the male and female *reproductive systems* describes the influence of the hormones on each, as well as the development of a new individual. The *lymphatic* and *immune systems* follow, with the comparison of the vessels to those previously covered and the anatomy and physiology of the system. The immune system covers the various mechanisms and actions conducted by the body to protect and prevent disorders and diseases. The last covered is the *endocrine system*, with description of the anatomy and physiology of its components and the impact on the organism.

LABELING ACTIVITY

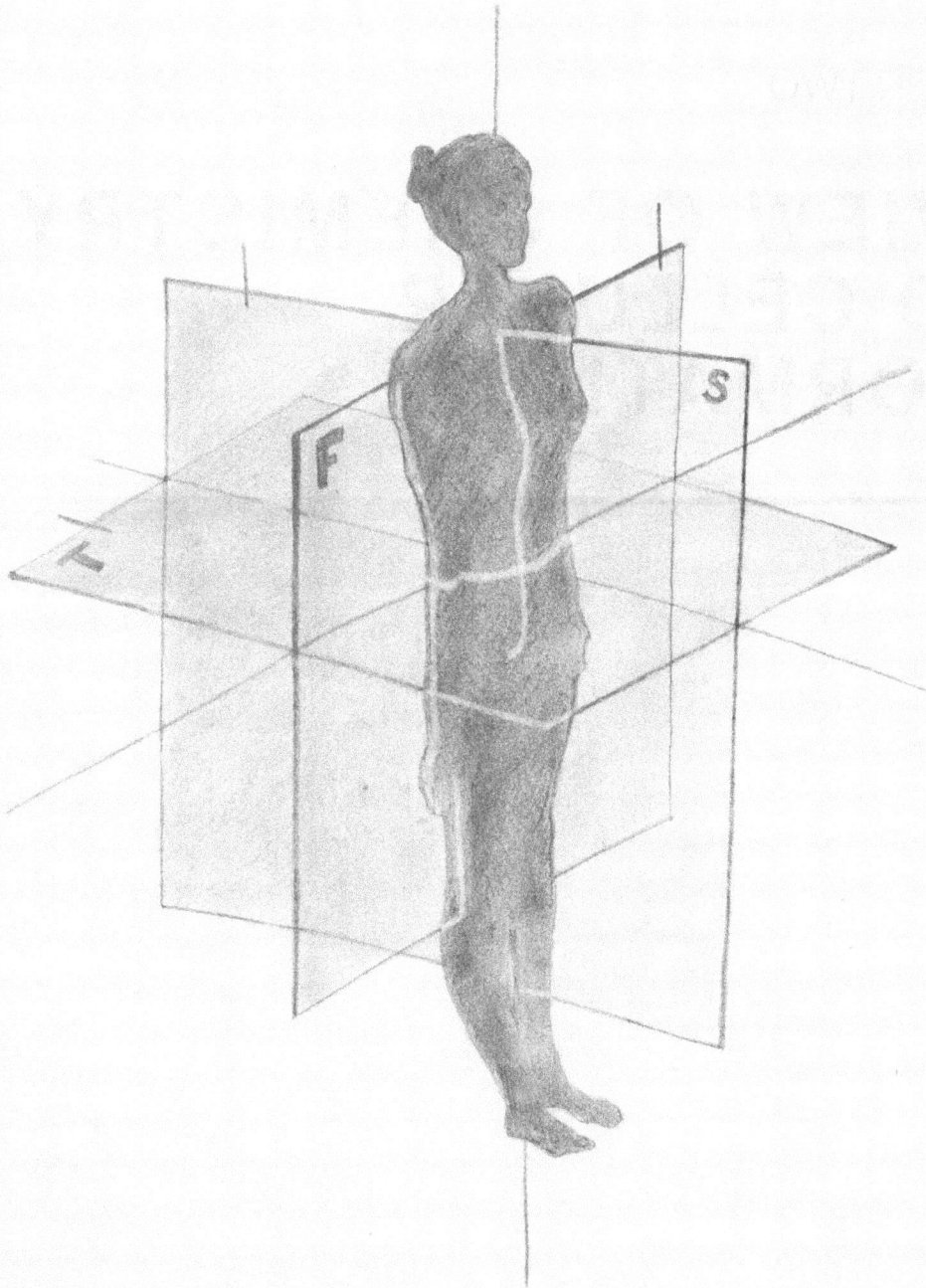

Fig. 1.1 Source: Copyright © 2013 Depositphotos/shotsstudio.

CHAPTER TWO

REVIEW OF CHEMISTRY AND CELLULAR INFORMATION

LEARNING OBJECTIVES

Upon completion of the chapter readers will have reviewed the essential knowledge of the cell and basic chemistry:

1. The organization of atoms and the role of each component in the formation of molecules
2. The understanding of pH and the role of buffers in homeostasis
3. The types of bonds found in molecules and the influence in solutions
4. The inorganic molecules, the classes of organic molecules, and their general function
5. The cell membrane and its role in transport
6. The organelles with the cell and their functions
7. The role of DNA and RNA in the production of ribosomes and proteins

Chemistry

The following information is a summary of essential concepts in chemistry that are critical in understanding fundamental life processes.

Elements (Atoms)

Matter is defined as anything that has mass and occupies space, which is also what defines an atom (element). *Elements* are the simplest substances which cannot be broken down by chemical means and have distinct physical and chemical properties. The majority of the human body is composed of only four elements (carbon, hydrogen, oxygen, and nitrogen), which represent ninety-six percent of our total body mass; when we include calcium and phosphorous, it increases to ninety-nine percent of the body's mass. Atoms consist of a *nucleus* containing *protons* and *neutrons*, with a cloud of *electrons* in orbits around the nucleus. Protons are positively charged, neutrons have no measurable charge, and electrons are negatively charged. Each element is assigned an atomic number that reflects the number of protons found in the nucleus and the number of electrons in the orbits. The protons and neutrons have a mass of approximately one *dalton*, which is reflected in the atomic mass of the element, with the difference from the atomic number being the number of neutrons found in the most common isotope of the element. When atoms interact in chemical reactions they form molecules, which result from the interactions with the outermost electrons, which are called the valence electrons.

Molecules

A *molecule* results from the chemical interaction that joins two or more atoms by chemical bonds. Compound is a term which may also be used to describe a molecule, but is defined by the atoms being of more than one type. In inorganic molecules the bond is the *ionic bond*, where the outermost electrons on one atom are attracted to the reacting atom, or the electron is lost to the reacting atom, to fulfill its outermost orbit with electrons. In many molecules the electrons are shared by the reacting atoms, this is common in organic molecules. These are *covalent bonds*, which frequently are stronger than ionic bonds. In biological systems, molecules interact due to the shape of the molecules having a positive or negative pole, so that weak bonds tend to tie the molecules to one another, which is common in proteins and nucleic acids.

Reactions

Chemical reactions involve the processes of bond breaking and rearrangement to form new molecules. When bonds are broken, energy is liberated and forms new bonds, which capture some of that energy in new molecules; some energy is lost as heat. There are two types of reactions: *synthesis reactions*, which form new molecules with a higher energy state, and *decomposition reactions*, where the new molecules formed have a lower energy state than the original molecules.

Inorganic Molecules

Most inorganic molecules are composed of ionic bonds and lack carbon. They tend to risk decomposition and are usually soluble in water to some extent. Oxygen, water, carbon dioxide, and ammonia are inorganic molecules that are covalently bonded; they are the exception to the ionic bonding characteristic. *Water* is the most common and abundant inorganic molecule, composed of hydrogen and oxygen covalently bonded. This polar molecule makes an excellent solvent, it traps and releases heat slowly, lubricates movable components of the body, and is involved in chemical reactions. Typical inorganic molecules are *salts*, which are the product of reactions between acids and bases. Salts dissociate in water, creating charged ions. If excess hydrogen ions are released then the solution will be acidic, and if excess of hydroxyl ions are released then the solution created is basic. Usually a solution is described by its *pH*, which expresses the concentration of hydrogen ions present in a scale from 0 to 14, which are the negative exponent of the base 10. Water in its pure state has a pH of 7.0 momentarily. A solution below 7 (0 to 6.99) would be an acidic solution whose strength is reduced the closer it gets to 7 and above (7.1 to 14); the strength of the basic solution increases the further it is from 7. Each whole unit difference is a ten-fold increase or decrease from the previous number. In biological systems extremes of pH are not tolerated, so there are a series of *buffers*, which trap and release hydrogen ions to maintain the hydrogen ion concentration within narrow limits so that successful activities can take place. In humans buffer systems exist in the blood, composed of carbonic acid in the plasma and hemoglobin in the red cells. In the cells, buffer systems involve phosphates and proteins all working together to maintain the slightly basic condition within the body.

Organic Molecules

All organic molecules are composed of carbon and hydrogen bonded by covalent bonds, creating molecules which are either polar molecules that are soluble in water or non-polar molecules, which are insoluble in water. Organic molecules with oxygen added to the carbon and hydrogen skeleton form carbohydrates and lipids.

Carbohydrates have a ratio of 1 carbon, 2 hydrogen, 1 oxygen (1:2:1) ratio. Sugars, starches, cellulose, and chitin are forms of carbohydrates and are polymers made up of various monomers by a dehydration synthesis. They can be broken into simple sugars by hydrolysis reactions, the removal or addition of water to the molecules. These reactions are driven by specific enzymes. Carbohydrates are the primary energy source, but can also form storage and structural components in the cells.

Lipids are another organic molecule to which oxygen is added to the carbon hydrogen skeleton, but lacks a fixed ratio. Lipids are non-polar covalently bonded and tend to be non-soluble in water. They consist of fats, steroids, phospholipids, vitamins, and prostaglandins, which insulate, protect, and store energy.

Proteins are polymers formed by the combining of various amino acids to create a molecule. They function in the protection, contraction, and regulation of structural components and enzymes. The amino acid is composed of a carbon skeleton that has two functional groups attached; one is an amino group containing nitrogen and hydrogen, and the other a carboxyl group containing a carbon with a double-bond oxygen and a hydroxyl group. Proteins are formed by ribosomes, in which a hydrogen is removed from the amino group, and a hydroxyl group from the carboxyl group, dehydrating the chain by creating the peptide bonds.

The last group of organic molecules is the *nucleic acids*, which are composed of a pentose sugar, nitrogenous group, and phosphate group. They function in the storage of information and the capture and transport of energy, and are involved in protein formation. DNA (deoxyribonucleic acid) is a double helix that stores the genetic information for the body. It is located in the nucleus of the cells. RNA (ribonucleic acid) transcribes the information from the DNA and translates it into proteins in the cytoplasm. Capture and transport of energy involves ATP (adenosine tri-phosphate), NADP (nicotinamide adenine dinucleotide phosphate), and FAD (flavin adenine dinucleotide) in various places to support life processes. Some trap and transport electrons and hydrogen, while others develop high-energy bonds to transport the energy.

Cell

The following is a summary of the essential cellular components and their function. Just as the atom is the smallest unit of matter, the cell is the smallest unit capable of carrying out the life processes. A *cytologist* is an individual who studies cells. Those at the technician level screen for atypical cells from various tissue samples. Cells are divided into those which have an organized nucleus surrounded by a nuclear membrane, and those which do not have an organized nucleus. *Prokaryotes* are those without an organized nucleus, and frequently lack membrane organelles. *Eukaryotes* have internal, organized organelles in addition to the nucleus. All cells, whether pro-karyotes or eukaryotes, have a membrane that surrounds the cellular contents called the *plasma membrane*.

Plasma Membrane

The plasma membrane is composed of a bi-layer of phospholipids, with the hydrophilic ends facing the interior of the cell or the environment. There are integral proteins embedded in the membrane, and the membrane is frequently stabilized by cholesterol. The configuration of the membrane makes it highly selective, permitting lipid-soluble materials to pass into the cell, con-trolling those molecules which are water soluble and polar passage into the cell. Most of the polar molecules that are transported into the cell are bound to carrier proteins or small enough to pass through pores in the membrane.

Movement across the Membrane

Movement of molecules across the plasma membrane is either active, where energy (ATP) is expended, or passive, where the cell expends no energy. *Passive transport* involves the kinetic energy of molecules or molecular size moving from higher to lower concentrations.

Diffusion is one form of passive transport that involves the kinetic movement of molecules from a higher to lower concentration location; this does not involve a membrane.

Osmosis is the movement of water from higher to lower concentrations through a selective membrane until equilibrium is established. A state where the solute concentration is lower inside the cell would indicate a *hypotonic condition* in the cell, with water moving out of the cell, attempting to reach equilibrium. If the concentration of the solute is equal on either side of the membrane, an *isotonic condition* exists in the cell, where water will move into and out of the cell equally. If the solute concentration is greater inside the cell than that surrounding the cell, water will move into the cell attempting to establish equilibrium. This is a *hypertonic condition*.

When a polar molecule such as glucose combines with an integral protein in the membrane, as the combination rotates with the fluid movement of the membrane the glucose is moved into the cell. This is *facilitated diffusion*, in which the molecular shape of the protein changes slightly, encouraging movement into the cell.

The last type of passive transport is triggered by the pressure exerted by water on the mem-brane. The pressure exerted by the water is the hydrostatic pressure and the process is *filtration*, where water is forced through membrane pores, removing other components in the process.

In *active transport* the cell expends energy (ATP) to move the molecules or components into the cell at time against a concentration gradient, an example would be ion pumps. In some cases, the energy is used to reconstruct the plasma membrane, such as in *endocytosis*, where the entrapment of extracellular materials occurs by surrounding them with extensions of the plasma membrane. There are two forms of endocytosis: *phagocytosis* is the ingestion of solid particles, which is used by white blood cells to ingest foreign bodies; *pinocytosis* is the ingestion of liquid materials. Receptor-mediated endocytosis is a form where large molecules are selectively ingested; an example would be the ingestion of antibody-labeled foreign antigen. The last of the active transports is the *ion-pump*. This is where integral proteins utilize the energy from ATP to move ions against a concentration gradient. Examples include the movement of sodium and potassium in muscle cells, and calcium and magnesium in the sarcoplasmic reticulum of skeletal muscle.

Cytosol

The cytosol is the cell contents between the plasma membrane and the nuclear membrane. This includes organelles, inclusions, and the cytoplasm (the fluid portion). It is composed mostly of water, which contains proteins, carbohydrates, and lipids as well as a variety of inorganic molecules. This is the location where many chemical reactions occur.

Organelles

In most cells, the *nucleus* is the largest visible organelle. It is a complex structure encapsulated by the nuclear membrane, which regulates the cell's activities and stores the genetic code the cell uses to form proteins and other cellular components. The nucleoplasm contains the chromosomes where the code is transcribed and a nucleolus where the ribosomes are formed, consisting of RNA and protein enzymes. Some tissues have multiple nuclei, because of the merging of cells during development such as that found in skeletal muscles, or lack a nucleus, in the case of red blood cells where the nucleus and mitochondria are lost as the cells mature.

Ribosomes are granular structures formed of ribosomal RNA and proteins transferred from the nucleolus to the cytoplasm, where they construct proteins from amino acids. Some are bound to the endoplasmic reticulum, creating a rough condition, and the majority of the ribosomes are free in the cytoplasm.

Endoplasmic reticulum is a network of parallel bi-layer membranes located in the cytoplasm that connects the plasma and nuclear membranes. It provides mechanical support and a site where materials are exchanged with the cytoplasm. Materials are transported throughout the cell in the endoplasmic reticulum. There are two distinct segments of the endoplasmic reticulum: one is the *smooth endoplasmic reticulum*, where many of the steroids are produced, and the second is the *rough endoplasmic reticulum*, where proteins are stored and lipoproteins are formed.

Adjacent to the endoplasmic reticulum is the *Golgi complex*, which is composed of eight stacked and flattened membrane sacs called cisternae. Many materials and molecules are processed and packaged in the Golgi complex for transport or storage within the cell. Some of these packages are the *lysosomes*, which are spherical vacuoles containing digestive enzymes. The lysosomes are involved in the digestion of phagocytized materials ingested by macrophages, and are responsible for cell destruction of worn out cells in a process called autophagy, where the cells undergo autolysis. Many lysosomes are secreted to extracellular sites to aid in digestion.

Mitochondria are the site where the majority of the ATP is produced in the cell by the citric acid cycle, which yields ATP, electrons, and hydrogen ions, which are transported to the electron transport chain and produce ATP by oxidative phosphorylation. The structure possesses a smooth outer membrane with folded cristae internally surrounding a matrix. The mitochondria has distinct ribosomes and genetic code separate from the rest of the cell.

The cytoskeleton of the cell is composed of various proteins that provide support, movement, and structural reinforcement of the cell. There are microfilaments, which consist of the proteins actin and myosin, which are involved in contraction, providing movement and support. Microtubules are slender tubes composed of the protein tubulin, which provides support and structure for the cilia, flagella, and centrioles, which form the mitotic spindle during mitosis. Intermediate filaments are composed of various proteins reinforcing the structure of the cell. In the resting cell the *centrioles* are paired cylinders arranged at right angles to each other in the dense *centrosome*. Extensions from the plasma membrane composed of tubulin are the flagella and cilia, which are used to promote motion. Cilia are commonly found in the trachea, reproductive ducts, and other locations, while the flagella are only found in human sperm.

Inclusions

Inclusions are any chemicals produced by the cell that are stored in the cytoplasm. Inclusions are usually organic and easily recognizable. Melanin is one type of inclusion, typically found in the epidermal cells as a brownish-black pigment that gives skin its color. Its production is accelerated with exposure to sunlight. Glycogen is another inclusion, which is animal starch used to store energy in the liver and skeletal muscles. The most obvious inclusion is the oil (lipids) stored in the specialized vacuoles in the fat cells.

Protein Production in the Cells

The code for the proteins is contained in the sequence of DNA, which first have to be transcribed. This occurs by the production of an RNA strand from the DNA. There are three forms of RNA produced. The first form is rRNA (ribosomal RNA), which is transported to the nucleolus to produce a ribosome. The second is mRNA (messenger RNA), which goes to the ribosome, initiates its activity, and defines the sequence of amino acids required to produce its protein. The last is tRNA (transfer RNA), which selects the appropriate amino acids and transports them to the ribosome to produce the protein.

Reproduction

Cells within the body are constantly reproducing; as some die, they are replaced by new cells. This process is by *mitosis,* which is the division of the nucleus through a series of steps developing replacement cells that are genetically and functionally identical. The cells then mature and produce additional DNA to form chromosomes during a period called *interphase* or *S phase.* The chromosomes are organized into duplicates during the *G2 phase* part of the interphase if the cell is stimulated to divide. The cells that are going to divide chromosomes condense, becoming visible while the nuclear membrane and nucleoli disappear and the centrioles migrate, forming poles during *prophase*. A series of spindles develop from the centrioles, with the chromosomes migrating and attaching to the spindles during *metaphase*. The chromosomes divide along a midline of the spindle and begin to be pulled toward the poles formed by part of the centriole during *anaphase*. As the chromosomes reach the poles, they tend to unwind and stretch out, becoming less apparent as a nuclear membrane begins to re-form around each pole, with the nucleoli reappearing during *telophase*. These are the stages of mitosis, which only defines the nuclear portion of cell division. The cell divides into daughter cells by a process called *cytokinesis*, which begins during the later stage of anaphase of the nuclear division and proceeds with the development of a plasma membrane along the midline of the dividing cell, until the cells separate into genetically identical daughter cells.

To produce a new organism, the chromosome number must be reduced to maintain organism characteristic functions. This is *meiosis*. Meiosis occurs as a two-stage division of the chromosomes, producing four genetically unique cells called *gametes*, which are incapable of survival unless fused with another gamete to obtain a complete genetic profile. The male gametes are *sperm* and the female gametes are *ova*, neither of which are capable of long-term existence unless preserved.

In some cases, cell division fails to be controlled; cell inhibition may be lost or the cell may divide rapidly, resulting in cancer cells, which can be called *neoplasms*. The loss of control can be correlated with a variety of viral and environmental factors that cause a modification in the genetic mechanisms that control growth and division of the cells.

CHAPTER THREE

HISTOLOGY

THE STUDY OF TISSUES

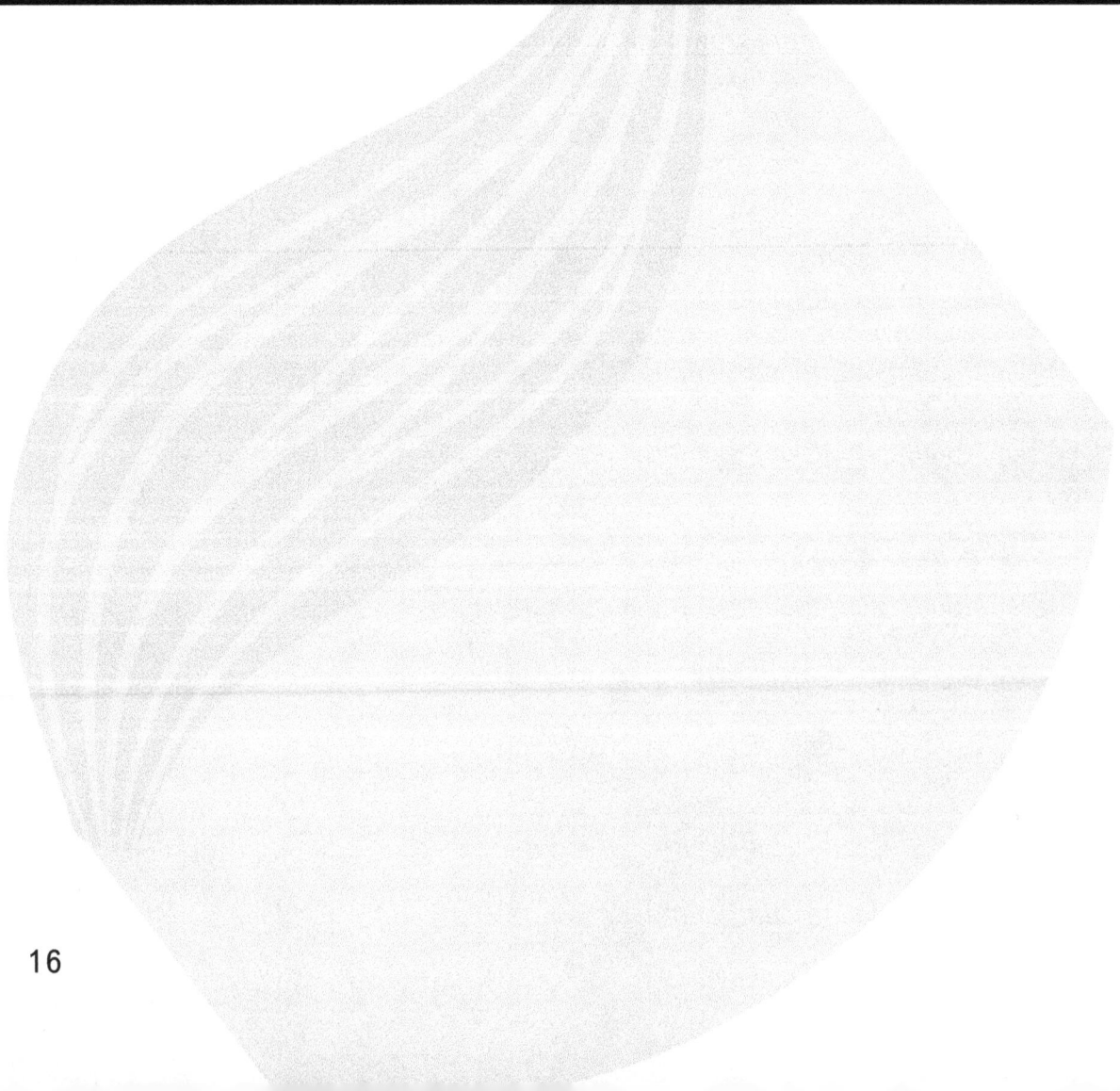

Upon completion of the chapter readers will have reviewed the essential knowledge of the cell and basic chemistry:

1. The description of the primary types of tissues and their functions
2. The composition of the matrices of each type of tissue
3. Detailed description of epithelial, connective, muscle, and nervous tissues
4. How repair occurs in damaged tissue
5. The anatomy and function of the types of membranes

Tissues are formed by groups of cells that are genetically alike, functioning to provide support of the body by the unique properties of the cell. The functional capacity of the tissue is greater than which would be expected by the performance of each individual cell. The study of the anatomy of tissues is *histology* and the study of the function would be *histophysiology*.

Tissues are groups of similar cells assembled together and bound by extracellular substances in the intracellular spaces to provide specialized activities required to support the life of the organism. Tissues are described and characterized by the cells and the associated quantity of the extracellular materials in the intracellular spaces. The combination of the cells and extracellular materials influence the function of the tissue and the locations where they are found.

The extracellular materials are referred to by a variety of names, such as interstitial fluid, plasma, and mucus. The most abundant extracellular materials are found in connective tissue and are frequently referred as matrix or ground substance. The matrix molecules support, strengthen, and bind embedded cells together and contribute unique properties to the tissues in which they are encountered. Examples of common extracellular materials include the following: 1) *hyaluronic acid*, which lubricates movable joints of the skeleton and aids in maintaining the shape of the eyeball; 2) *chondroitin sulfate*, which provides support and adhesion of cells in cartilage, bone, heart valves, umbilical cord, and the cornea of the eye; 3) *collagen fibers*, which are found throughout cartilage, bone, and skin in varying concentrations and make up the majority of tendons and ligaments, being inelastic and providing strength and support; 4) *reticular fibers*, composed of small strands of collagen wrapped with glycoproteins providing the support network around fat and nerve cells, wrapping around muscle cells, and elasticity to blood vessels and the skin; and 5) *elastic fibers*, composed of the protein *elastin*, which provide flexibility and support to the skin and blood vessels.

Epithelial Tissue

Epithelial tissues are defined as the tissues that cover or line spaces. This includes the skin and various membranes lining spaces around the heart or abdomen where it lines the space and covers the internal organ such as the pleural membrane. The cells of the epithelial tissue on the outermost location have a surface that does not contact any other cell; this is called a free surface. Cells in epithelial tissue are supported by a *basement membrane*, formed by the underlying connective tissue interacting with the cells of the epithelial layer. When you name epithelial tissue, the first characteristic is the shape of the cells: those which are flattened are identified as *squamous*; those which resemble a square box are identified as *cuboidal*; those which are elongated are identified as *columnar*; and those which change shape under tension are *transitional*. The second characteristic used to name epithelial tissue is the arrangement of the cells in layers: if the cells are all in a single layer, the tissue is named *simple*; if the cells are in two or more layers, the tissue is named *stratified*; and if mature and immature cells are grouped together, appearing as though a layer exists, the tissue is named *pseudostratified*. Little extracellular material is present in epithelial cells, which are attached tightly together, providing the tissues with unique abilities. Epithelial tissues do not have a direct blood supply, which makes them *avascular*. Additional names associated with epithelial tissue are *mesothelial*, which are tissues forming some middle layer (usually membranes) in closed body cavities, and *endothelial*, which line internal structures like the blood vessels or the heart.

The epithelial tissues which line or cover the body have unique properties dependent upon the character of the tissue. *Simple squamous epithelial tissue* is found in the alveoli of the lungs, lining the blood vessels and forming the capillaries, where they function to diffuse gases and move fluids and dissolved substances by osmosis at a variety of locations in the body. *Simple cuboidal epithelial tissues* are found in the salivary glands and tear ducts, where they are involved in secretion, and lining the tubules in the kidney, where they are involved in the absorption of water from the filtrate. *Simple columnar epithelial tissues* are abundant in the digestive tract, where they are involved in the absorption of nutrients and have the free surface area of the cell convoluted into microvilli to increase surface area. *Stratified squamous epithelial tissues* are found in the skin and openings to the exterior, where they provide protection against microbes; some are filled with a waxy protein called keratin and others are very moist. *Stratified cuboidal epithelial tissues* are found in the conjunctiva of the eye, the urethra of the male, and sweat glands. They function in the protection of the structures where they are located. *Stratified columnar epithelial tissues* are found in the male urethra, where they provide protection and secretion. *Pseudo-stratified columnar epithelial tissues* are found in large excretory ducts and in the respiratory tract, where they provide protection and movement of material along a tract. *Transitional epithelial tissues* are found in various bladders, where they retain fluids and distend.

Connective Tissues

Connective tissue is the most diverse and abundant type of tissue found in the body. It is composed of cells surrounded with an abundance of extracellular materials. Most connective tissue is vascular, with the exception of cartilage. It is involved in protecting, supporting, and binding elements of the body together, and in storing energy. Embryonic connective tissues are the mesenchyme and mucous connective tissue, which are found in the embryo and later in the umbilical cord. Adult connective tissues are diverse, having various cells contained in a matrix of differing characteristics.

The first connective tissues to be considered are those grouped as loose, where the cells are not in a fixed position. These are *areolar tissue*, located in organs, abundant in the dermis and underlying layers; it consists of fibroblast, macrophages, and plasma cells contained in a matrix containing collagen, elastic, and reticular fibers in a jelly-like mass. *Adipose tissue* is another type of loose connective tissue, located throughout the body in the dermis, around the heart and kidneys, and in the yellow bone marrow and surrounding joints, where it is composed of storage cells in a fibrous matrix of reticular fibers, providing energy storage, support, and protection of vital structures.

Another grouping of connective tissue is *collagenous connective tissues (dense)*, which are composed of fibroblasts that produce an abundance of extracellular collagen, which dominates the tissues; these are found in tendons (regular type) and ligaments (irregular), providing the heart, liver, kidneys, and lymph nodes structural support and protection throughout. Another *collagenous connective tissue (elastic)* is found in the vocal cords, lungs, and trachea. The amount of collagen is reduced and replaced by elastic fibers, providing flexibility and stretching. A third type of *collagenous connective tissue (reticular)* is found in the structure (stroma) of the liver, spleen, lymph nodes, and the basal lamina.

The next group of connective tissue is cartilaginous connective tissues, which have a large amount of chondroitin sulfate and glucosamine in the extracellular matrix surrounding scattered cells (*chondrocytes*), located in *lacuna* or spaces. These tissues lack a blood supply. The first type is *hyaline cartilage*, which is found covering the ends of long bones, attaching the ribs to the sternum, providing support to the trachea and bronchi, and forming the structure of the larynx, as well as forming the embryonic skeleton; it provides flexibility, support, and movement. The second type is *fibro-cartilage*, which is found in the knee as the menisci, in the vertebral column as intervertebral discs, and as the attachment at the pubic symphysis; the matrix is filled with an abundance of collagen fibers that function to absorb energy, provide support, and fuse bone components together. The third type is *elastic cartilage*, which is found in the epiglottis, Eustachian tubes, and the external ear; collagen fibers are replaced in the matrix by elastic fibers, providing flexibility.

The next type of connective tissue is composed of a dense, mineral-rich matrix surrounded by a thin fibrous covering; this is the *osseous connective tissue (bone)*. The cells (*osteocytes*) are located in organized structures in the mineral matrix or in the *periosteum* covering the bone. The mineral salts making up the matrix are primarily calcium and phosphorous, supported by collagen fibers. The mineral salts provide hardness, while the collagen provides elasticity and strength. The bone provides protection to vital structures, supports various structures, stores minerals, provides movement, and houses blood-forming tissues.

The last type of connective tissue has a fluid matrix in which diversified cells and components are located; this is *vascular connective tissue (blood)*. Blood is composed of a liquid matrix (*plasma*) composed mainly of water, and contains salts and nutrients as well as cellular components (*platelets*) and either red blood cells or white blood cells. The various components of the blood are involved in transport, removal of pathogens or cellular debris, immune reactions, and prevention of blood loss (clotting).

Membranes

Membranes are classified as epithelial membranes because they are composed of an epithelial layer, which is the functional area and is supported by an underlying connective layer. There four distinct membranes found in the body:

1. The *mucus (mucosa) membrane*, which is moist, lines openings to the exterior of all body cavities; they provide a barrier to pathogens and flush pathogens from the surface, preventing adherence while trapping foreign particles in the *mucus* secreted, with the layer between the epithelial and connective layers forming the *lamina propria*.
2. The *serous membrane (serosa)* is the membrane found in closed body cavities. It is divided into a portion that covers the internal organs, named the *visceral* portion, and the lining of the surrounding cavity, named the *parietal* portion, with a liquid being secreted between the portions named the *serous fluid*, which lubricates the movement of the internal structures. Examples of serous membranes are the pleural and pericardial membranes in the thoracic cavity and the peritoneal membrane found in the abdomen.
3. The *cutaneous membrane* (skin, integument) is the largest of the epithelial membranes, composed of keratinized epithelial cells and diverse connective support tissues separated by a basement membrane.
4. The *synovial membrane* is unique in the epithelial membranes in that it is dominated by the connective tissue, with small islands of epithelial tissue that secrete *synovial fluid* to lubricate the articulations of the skeleton. The synovial membrane is found in tendon sheaths, joint cavities and in *bursae*, aiding in the movement and flexibility of the skeletal components.

Muscle Tissue

Muscle tissue consists of cells that contain organized strands of cytoskeletal proteins (*actin* and *myosin*) that have the ability to contract when stimulated. When these proteins contract they exert tension, which transfers as force to some point. Muscles provide motion in the body, maintain posture, and generate the majority of the heat the body requires for homoeostasis. There are three distinct types of muscle tissue.

The *skeletal muscle tissue* is the most abundant and extensive, with all being attached directly or indirectly to the skeleton. The original cells merge during embryonic life, creating muscle fibers that are multi-nucleated, with the nuclei forced to the sides of the fiber; the fibers are elongated and have cross bands called striations that reflect the *sarcomere* structure of the actin and myosin. These muscle fibers are capable of generating high force levels under control (quickly or slowly) as required by conscious control of the contraction of the muscle. They are capable meeting energy needs aerobically and anaerobically, but the endurance is limited, as they easily fatigue. Some skeletal muscles contain stored oxygen in *myoglobin* and all fibers store *glycogen*.

Cardiac muscle (myocardium), which is only found in the heart, possesses characteristics similar to skeletal muscle and smooth muscle. Like skeletal muscle, it appears striated, with the contracting elements being organized into sarcomeres. Like smooth muscle, cardiac muscle has a single centrally located nucleus. Unique characteristics of the cardiac muscle are that it is highly branched, being in contact with other cardiac muscle cells with a tight gap junction; the *intercalated disc*, which aids in the spread of the stimulation-triggering contractions; in an involuntary rhythmic fashion (wave-like). The contracting stimulus is generated internal to the heart. This muscle has numerous mitochondria and stores oxygen in myoglobin to produce the energy aerobically for contraction.

Smooth (visceral) muscle is the muscle tissue found in internal structures used to provide motion, change lumens, control bladders, and modify blood pressure. The cells have a central nucleus, lack striations, and contract slowly, so they tend not to become fatigued. The tissue is organized into layers, with all the muscles cells in the layer oriented in the same direction. They are involuntary with the cells responding to local, generalized, specific, and chemical stimulations.

Nervous Tissue

Nervous tissues are composed of two types of cells that are functionally and anatomically different from each other. The *neuron* cells are electrically excitable, having the ability to respond to a stimulus and generate an action potential to transmit the information to another location. Neurons are composed of a *cell body (soma, perikaryon)* containing the nucleus and two processes extending from the cell body: *dendrites*, which bring action potentials toward the cell body, and *axons*, which carry action potentials to other locations. Structurally, there are three types of neurons—multipolar, bipolar, and unipolar—while functionally there are those which are sensory, motor, or interneurons. Neurons are incapable of regeneration. The supporting cells in the nervous system are the *neuroglia (glial cells)*, which perform the function of connective tissues. Glia are capable of regeneration. There are four types found in the brain and spinal cord, and two types found in the rest of the body.

Tissue Repair

Any organ or tissue can be damaged, requiring repair or replacement to restore the body function called healing. In organs, the tissue that carries out the function of the organ is named the *parenchyma*, composed usually of epithelial tissue, most of which is capable of regeneration. The *stroma* is the portion of the organ that provides support and shape, and is usually composed of connective tissue capable of regeneration. To restore the organ's function, the parenchyma tissue must regenerate; if the parenchyma is unable to regenerate, the tissues of the stroma will replace the functional tissue, restoring the organ's shape. This is frequently done by fibroblasts creating collagen and granulations (scars) or adipose tissue. When multiple layers of tissue are damaged, such as in surgery, healing and the development of scarring can tie together layers that usually are free of each other, creating *adhesions*. Healing of damaged tissue requires good nutrition with adequate protein and vitamins, and abundant local blood supply, along with growth hormones. Age slows the healing process, due to available growth hormones to stimulate the regeneration of the tissues

CHAPTER FOUR

THE INTEGUMENT SYSTEM

ORGANIZATION AND FUNCTION OF THE CUTANEOUS MEMBRANE

LEARNING OBJECTIVES

Upon the completion of the chapter readers will have an essential knowledge of integuments anatomy and physiology:

1. The role of the integument in maintenance of homeostatsis
2. Organization structures contained in the integument
3. The physiology of the epidermis and dermis
4. The accessory structures located in the integument anatomy and function

Introduction

The integument is the first complete organ system to be addressed in this course. It extends over the exterior of the body and is approximately nine square meters of the surface area of the average individual. It provides protection from the surrounding environment, and senses the environment at the same time. The intact integument prevents the loss of water from the body and gain of water from the environment. Microbes are inhibited on the integument by the composition of the contents of the surface cells and oils released on the surface. The medical professional whose specializes in the diagnosis and treatment of the integument is the *dermatologist*. The integument reflects the health and condition of the individual, as it is affected by nutrition, hygiene, circulatory issues, allergic stresses, genetic traits, psychological state, and medications being taken. The integument consists of an outer layer of epithelial tissue, the epidermis, and an inner layer of a variety of connective and other tissues, the dermis.

Anatomy and Physiology of the Integument

The *epidermis* is the thinner layer of the skin, composed of epithelial cells that are generated near the dermis and, as they progress to the surface, produce keratin, a fibrous waxy protein that provides toughness to the epidermis and aids in physical protection. Once these cells reach the surface, they are filled with keratin and they die.

The *dermis* is the thicker layer, composed of connective, nervous, and epithelial tissues. It provides nutrients and waste removal from the epidermal cells, enabling the production of replacement cells for the epidermis.

The functions (physiology) of the integument aid in ensuring life processes are sustained.

1. The skin aides in the regulation of the body's temperature (thermoregulation) by sweating to release heat from the body, trapping air on the surface to reduce heat loss by aiding the

adipose-rich layers to insulate the body, and shunting blood away from the skin to reduce radiant heat loss.

2. Physical and chemical protection of the body is provided by the epidermal cells at the surface of the skin. The oils released from the hair and sweat create an unfavorable environment for most microbes.

3. The skin acts as a sensory organ, detecting gentle movement through the hairs, light pressure through sensors in the upper dermis, heavy pressure through sensors deep in the dermis, environmental temperatures through thermal sensors, and pain through free nerve endings.

4. Excretion of waste organics and electrolytes occur in sweating and oil release, which aids in the reducing the microbial populations on the surface.

5. The skin is the first line of immune response, with the Langerhans cell in the epidermis triggering response to contact allergens.

6. At any time at rest the integument will contain approximately twenty percent of the body's total blood supply; this provides a reservoir of blood that can be relocated as a need arises, or suddenly the volume of blood in the integument may increase which results in blushing or becoming pale as blood departs the skin during stress.

7. The skin is the site where the sunlight promotes the conversion of lipids to the precursor of vitamin D_3 when exposed to at least one hour per week. This precursor is then activated in the liver and kidneys, promoting the absorption of calcium in the small intestine.

Epidermis

Keratinocytes are the most abundant cells in the epidermis. About eight percent of the cells are *melanocytes* which produce melanin, absorbing ultraviolet light energy, reducing damage to the rapidly dividing keratinocytes. *Langerhans cells* trigger response to contact allergens and *Merkel cells* are sensitive to light touch. Dependent upon the location in the body, there are four or five distinct layers in the epidermis. Generation of the epithelial cells of the epidermis occurs in the *stratum basale* (*germinativum*), which is a single layer of rapidly dividing cells atop the dermis, and separated from the dermis by the basement membrane that permits the diffusion of nutrients from the dermis to support the needs of the dividing layer. Once the cells divide they move into the *stratum spinosum*, which is about ten cells thick, to mature, increase in size, and begin to flatten as they progress toward the surface. These mature cells form the *stratum granulosum*, which is only three to five cells, thick where they become completely filled with keratin and lose the ability to continue metabolic activities. In the palms of the hands and the soles of the feet is the *stratum lucidum*, which is eight to ten cells thick and provides additional protection for nerves, blood vessels, and articulation from the pressure to which the feet and hands are exposed, forming the thick skin. The outermost layer of the epidermis is the *stratum corneum*, which is twenty-five to thirty cells thick, usually thinner on the anterior body, and consists of cells that are filled with keratin and have died. The process of the epidermis cells requires between two and four weeks for the cells to proceed from the division in the stratum basale to reach the surface and be lost to the environment through abrasion. During this interval the cells become filled with keratin through a process called *keratinization*, which disrupts the cells' ability to carry out metabolic processes, leading to the death of the cells.

Dermis

The dermis is composed of diverse connective tissues containing collagen, elastic and reticular fibers, fibroblasts, macrophages, and areolar and adipose connective tissue. Hair follicles and sweat glands are organs composed of epithelial tissues and sensory nervous receptors—*Meissner corpuscles*, sense light touch; *Ruffini corpuscles*, sense heavy continuous touch; and *Pacinian corpuscles* sense deep pressure—are dispersed throughout the dermis. The region of the dermis closest to the epidermis consists of a convoluted area composed of areolar tissue, Meissner corpuscles, and capillaries, which provide nutrients for the epidermis by diffusion through the basement membrane. This region is called the *papillary region* or layer, due to the appearance of having little hills increasing the surface contact with the epidermis. The majority of the dermis is called the *reticular region* or layer. It is composed of irregular connective tissues containing collagen and elastic fibers, adipose tissue, and hair follicles, to which the *sebaceous* (oil) *glands* are attached, along with the *arrector pili* (smooth) muscle and the *sudoriferous* (sweat) *glands*. The dermis provides strength, extensibility, and elasticity to the integument due to its fibers. In the dermis, because of the arrangement of the fibers, lines of tension (cleavage) are created. Understanding these lines of tension allows surgeons to minimize scarring following injury or surgery. The papilla of the papillary region form epidermal ridges that create distinct patterns, which become obvious in areas of thick skin such as on the fingers, hands, and feet.

Organs of the Integument and Other Factors

The coloration of the skin is controlled by genetics fixed by the survival of our ancestors. Those whose ancestors were exposed to high-intensity ultraviolet light fixed the ability to produce the large quantities of melanin found in those who are dark skinned. Those whose ancestors were exposed to low levels of ultraviolet light produced less melanin and, permitting the reflection of hemoglobin in the dermis, are the pinkish or light skinned. Those whose ancestors concentrated carotene have skin coloration reflecting the yellow-orange. Everyone possesses a similar number of melanocytes, which are stimulated by the environment's ultraviolet intensity stimulates these melanocytes to produce our skin tones in combination the other factors.

Hair Follicles

Hair follicles develop from the embryonic epidermis. Development begins over a dermal papilla, the epidermis develops above it, and it moves into the dermis in the adult. The hair shaft provides sensation and protection by reducing heat loss combined with other components coupled with the follicle. The shaft is the portion growing above the skin. The follicle is the portion located in the dermis, and where the cells that form the shaft are produced from epithelial layers similar to those of the epidermis. Attached at the base of the follicle is a papilla containing nerve endings and capillaries. The follicle obtains its nutrients from the capillary bed, which provide the energy to produce the cells that will mature, forming the hair shaft. Attached to the portion of the follicle extending toward the surface are the sebaceous gland and arrector pili muscle. The sebaceous gland produces *sebum*, which lubricates the hair shaft, maintaining its flexibility and carrying sebum to the surface, aiding the epidermis. The arrector pili muscle responds to stress and temperature, pulling on the follicle and forcing the shaft upward, creating "goose bumps." The color of hair is due to its melanin content and sebum contained in the shaft. Hair grows in a cyclic pattern due to hormones and other stimuli, which also determine the patterns and texture.

Glands of the Dermis

There are four types of glands associated with the dermis, some being widespread and others found in only limited area.

1. The *sebaceous glands* are associated with the hair follicles and are only found where the body has hair. They are totally absent in the palms of the hand and the soles of the feet. The sebum acts to moisten the hair and skin, aid in waterproofing the epidermis, soften the epidermis to improve its flexibility, and inhibit microbial growth by the pH of the fatty acids contained in the sebum.
2. The *sudoriferous glands* are located throughout the dermis over the body. They have ducts that carry the sweat accumulated from the dermis to either the epidermis, which are called *eccrine sweat glands,* or into the hair follicles in the axilla and other locations, which are called *apocrine sweat glands.* The production of sweat is controlled by body temperature or the heat being relocated to the dermis, and the fluid contains water, small amounts of urea, and salts. Even when we are unaware we will still produce sweat.
3. The *ceruminous glands* are modified apocrine sweat glands that form a waxy mass with epidermal cells called *cerumen*, and are located in the external ear canal. The cerumen provides protection to the ear drum, keeping it pliable and waterproof, inhibits microbes, and entangles other foreign organisms.
4. The mammary glands are considered to be specialized sudoriferous glands located in the breast area, which are stimulated by sex hormones and others to produce specialized fluids.

Nails

The nails found on the posterior aspect of the fingers and dorsal aspect of the toes are composed of clear, hard keratinized epidermal cells, covering nerve endings and capillary beds. The nails aid in the grasping and manipulation of small objects. Some suggest that the presence of nails amplify the sensitivity of touch at the ends of the fingers.

CHAPTER FIVE

THE SKELETAL SYSTEM

ROLE IN SUPPORT, FUNCTION, AND PROTECTION OF THE HUMAN

LEARNING OBJECTIVES

Upon completion of the chapter readers will have the essential knowledge of the skeletal system and its articulations:

1. The description of types of bones and their location
2. The hormonal regulation of minerals in the skeletal system and body
3. The description and function of the long bones
4. How compact bone is repaired
5. The axial skeleton, its bones, divisions, articulation, and land marks
6. The description of vertebrae and the development of the vertebral column and its divisions and articulations
7. Appendicular skeleton division girdles, bone of the appendages, articulations, and movement
8. Specific description of all types of articulations, locations encountered, and their function

The skeletal system is composed primarily of connective tissues, with the osseous tissue dominating the system. Protection and movement are the major functions of this system.

Functions

The skeleton forms the framework of the human body, supporting internal organs, protecting vital organs, and promoting movement of the individual. Soft tissues are supported and muscles are attached to the bone, creating the framework of the body. Internal organs are enclosed by bone, providing physical protection to the more critical organs such as the brain, spinal cord, heart, and lungs. Movement is supported and enabled by the creation of leverage from the muscle contractions and articulations. The skeleton acts as a reservoir of mineral calcium and phosphorous, aiding in the maintenance of mineral homoeostasis. Energy is stored in yellow marrow found in the long bones. Bone provides the protection and site for those cells that produce blood cells in the red marrow, at the end of long bones and in other locations throughout the body.

Anatomy

Bone is composed of a mineral matrix composed of calcium salts (*calcium carbonate*, calcium phosphorous complex (*hydroxyapatite*) reinforced by a matrix of collagen fibers produced by *osteocytes*. The bone's hardness is provided by the mineral salts, and the tensile strength is provided by the collagen fibers. The process of depositing the mineral is called ossification or mineralization.

Cartilage is another component of the skeletal system. It provides the original architecture for the body. Hyaline cartilage blocks in the embryo layout the original skeleton, which are replaced by bone. Hyaline cartilage in the adult covers the ends of the long bones (articular surfaces) and attaching the ribs to the sternum. Fibro-cartilage is found in the adult skeleton, supporting and fusing the vertebrae as the intervertebral discs, the pelvic bones anteriorly as the pubic symphysis, and as the menisci of the knee.

The *periosteum* is a thin fibrous membrane covering the surface of bone and is absent on the joint surfaces. It contains osteocytes and stabilizes the attachment of nerves and blood vessels to the bone. It is active in the repair and remodeling of the bones throughout life.

Types of Bone

Compact bone is composed of dense deposits of minerals with the osteocytes scattered in lacunae in the matrix. In thicker areas of compact bone, they are organized into *osteonic systems*, which ensures that the osteocytes are provided with nutrients and the wastes are removed. The osteonic system consists of a central canal containing blood vessels and nerves, surrounded by a series of layers (lamella) containing the lacunae of the osteocytes. Connected to the central canal are a series of smaller canals (canaliculi) that permit diffusion of nutrients and wastes. As the compact bone remodels these osteons, continual change creating a series of patterns that appear like a series of bull's eye targets—some are complete and some are partial. The shaft of the long bones, such as the humerus or femur, is composed of compact bone. In other areas of the skeleton, thin layers of compact bone form the surface, and the osteons do not develop covering spongy bone such as in the flat bones of the skull. The dense structure of compact bone makes it resistant to injury by compression along the long axis, but makes it prone to fracture by shear forces.

Spongy (cancellous) bone forms the internal layers at the ends of long bones and flat bones. The osteonic systems are absent, as these areas contain the red marrow. Spongy bone consists of *trabecular spaces* housing the marrow. The lattice structure tends to disperse force over a wide area, absorbing the energy of jumping and impacts, thus reducing the likelihood of fracture. Spongy bone is prone to crush-type injury.

Bones Grouped by Length and Shape

Bone characterized as *long bones* consist of a shaft (*diaphysis*) composed of compact bone, and on both ends the *epiphysis* composed of spongy bone housing red marrow. A transitional area between these is sometimes identified as the *metaphysis*, containing the growth plate where the bones lengthen, identified as the *epiphyseal plate*. Covering the epiphysis is articular cartilage, hyaline cartilage that permits smooth movement of the articulation. The outer surface of the bone is covered by the periosteum extending over all but the articular surfaces. In the diaphysis is found a *medullary (marrow) cavity* containing yellow marrow, rich in stored fat. Examples of long bones are the femur, radius, humerus, ulna, tibia, and fibula.

Carpals are identified as *short bones*, which have a thin covering of compact bone covering a small amount of spongy bone. Many of the bones of the skull are identified as *flat bones*, which have a surface layer of compact bone over spongy bone. The vertebrae are identified as *irregular bones*, with the body being spongy bone and the processes and area over the spinal cord being compact bone. *Sesamoid bone* has a lazy S shape covering articulations, the largest being the *patella (knee cap)*.

Formation of Bone: Embryology

The embryonic skeleton is formed as either fibrous membranes or blocks of hyaline cartilage. The skeleton of the embryo begins the replacement of these during the sixth or seventh week of gestational life, and the process of ossification continues throughout life as the skeleton dynamically remodeling in response to life. The replacement of the fibrous membranes is *intramembranous ossification*, which can be seen in the flat bones of the skull of a newborn, which are called soft spots, fontanels. The replacement of the cartilage by bone is *endochondral (intracartilaginous) ossification*, where the bone begins to form around the cartilage blocks and the cartilage degenerates, creating the marrow cavity of the long bones.

Homoeostasis, Growth, and Repair of Bones

The calcium and phosphorous content of bone is dynamic, based on the concentration of calcium in the blood. If there is inadequate calcium in the blood, *parathyroid hormone* stimulates the destruction of hydroxyapatite, making calcium available for the blood, and promotes the loss of phosphate through the kidneys. If blood calcium is high *calcitonin*, a hormone produced by the thyroid gland, stimulates the development of calcium carbonate or hydroxyapatite, depositing calcium in the bones and maintaining calcium homoeostasis. Age impacts the skeletal system by loss of minerals in the bone because of reduced activity and hormone levels, particularly the reduction of levels of sex hormones. In addition, dietary changes occur due to changes in tastes, with individuals consuming less protein and more carbohydrates. Coupled together, these lead to bones with less hardness, the shrinkage of stature by the effects of *osteoporosis*, and changes in tensile strength, leading to embrittlement that increases the risk of fractures.

The lengthening of long bones begins before birth and continues until the mid-twenties, in most individuals ceasing in the late teens. The elongation of long bones occur from the epiphyseal plate, which has a series of zones of activity. The *zone of resting cartilage* is where the cells that have divided are preparing for the next division. Once division of the cartilage begins, it is termed the *zone of proliferation*. These cells grow and mature, then cease growing. Once the cells have ceased to grow they enter the *zone of hypertrophy*. Shortly after, the cartilage undergoes calcification, replacing the cartilage with minerals, becoming bone.

Bones grow in diameter by the action of *osteoblasts* (osteocytes that build bone) located in the periosteum. As stress is applied to the periosteum by muscle activity, this promotes increased blood movement in the area of stress, making available more resources. This promotes the formation of more collagen fibers and the deposit of more minerals. If the stress is removed for a long period of time, the bone will remodel itself, removing minerals and collagen. This is constant throughout life as old bone is replaced by new bone. The old bone is broken down by the action of *osteoclasts* (cells found in bone that are capable of demineralizing bone and decomposing collagen). New bone is then produced from the recycled minerals and proteins by *osteoblasts*. The activity of remodeling and growth is dependent upon having adequate minerals available, specifically calcium, phosphorous, magnesium, manganese, and zinc. Vitamins are essential to support the processes as cofactors, specifically A, B_{12}, C, and D. Hormones are essential to stimulate and support the processes, specifically *human growth hormone*, *insulin*, thyroid hormones, *parathyroid hormone*, and *calcitonin*. Moderate exercise promotes bone remodeling, creating stress on the periosteum.

Breaks (fractures) occur with stress overload of the bone at the break site. Frequently, the blood vessels in the periosteum and in the bone are broken, leading to bleeding in the area of the break. A clot forms usually within the first few minutes or the first twelve hours after the trauma, which is called the *fracture hematoma*. Over the next twenty-four to forty-eight hours, fibroblasts are attracted to the site by chemicals released from the clot and damaged tissues. The fibroblasts produce collagen fibers to begin to knit and stabilize the fracture site. This is called the *procallus* stage, where granulation begins. As the site becomes stable, fibro-cartilage begins to form, creating the *soft callus* stage, which occurs between forty-eight hours and about a week following the break, lasting about three to four weeks, dependent upon the age of the individual. The fibro-cartilage is mineralized between four to six weeks following the injury, creating a *hard callus*. The injury site will continue to remodel over the next several months and years. A fracture site will leave indications on the bone throughout life.

Axial Skeleton

The axial skeleton is composed of approximately eighty bones. It consists of long, short, flat, and irregular bones arranged along the longitudinal axis of the body, organized as the skull, hyoid, vertebral column, sternum, and ribs.

Skull

The skull is composed of twenty-two bones that fall into one of two divisions: the *cranium*, composed of eight bones, and the *facial*, composed of fourteen bones. The skull contains a series of cavities that lighten it, aid in the insulating of the brain, and aid in warming and moistening the air we breathe. These cavities are called the *paranasal sinuses*. These are located in the maxilla bones of the face, frontal, ethmoid, and sphenoid bones of the cranium. The *maxillary sinus* is always open to the exterior; the sinus located in the frontal bone (*frontal sinus*) may or may not be open to the exterior. Those found in the ethmoid (*ethmoidal sinuses*) and sphenoid (*sphenoidal sinuses*) are rarely open to the exterior. The bones of the skull articulate with each other through a series of *sutures*, which are fibrous, immovable (synarthroses) articulations. These sutures mark the boundaries where the fibrous membranes that formed in the embryo existed and mineralized, forming the flat bones.

The *cranium* is composed in the adult of a single *frontal*, paired *parietal*, *occipital*, paired *temporal*, *ethmoid*, and *sphenoid* bones. The frontal bone is joined to the parietal bones by the *frontal (coronal) suture*. The parietal bones are joined together by the *sagittal suture*. The parietals are joined to the occipital by the *lambdoid suture*. The parietals are joined to the temporal bones by the *squamous sutures*. The ethmoid and sphenoid bones are attached to one another and to other bones by various sutures whose names are not usually considered. The four named sutures represent the locations of the fetal and infant fibrous membrane boundaries called *fontanels*, which create the soft spots that allow the brain to grow and expand, remaining flexible and providing growth margins until they ossify between the ages of six and ten years.

Facial division of the skull in the adult is composed of fourteen bones. Most of the facial bones are paired, with the exception of the vomer and mandible. The facial bones are as follows: *nasal bones* along the bridge of the nose; *lacrimal bones* inside the orbit of the eye near the nose; *maxilla*, making up the central portion of the face; *zygomatic bones*, making up the upper and outer portions of the cheek; *mandible*, making up the lower jaw, *palatine bones*, making up the posterior third of the hard palate; the *vomer*, making up the septum of the nasal passages; and the *inferior nasal conchae*, which support the mucus membrane of the nasal passages. All of the facial bones, with the exception of the mandible, are joined by sutures. The mandible articulates with the temporal bone in a freely movable, condylar articulation identified as the *temporomandibular articulation*.

A specialized structure of the skull is the orbit of the eye. This is composed of seven bones normally, with an occasional eighth bone being the nasal. The primary bones considered to be in the orbit are the ethmoid, frontal, and sphenoid from the cranial; and the maxilla, zygomatic, lacrimal, and palatine bones from the facial.

The *hyoid* is a unique bone that does not articulate with any other bone. It is U-shaped, located in the neck region, attached to the larynx (voice box) and suspended in the musculature supporting the tongue. It is attached to the styloid process of the temporal bone and the upper vertebrae (cervical).

Vertebral Column

In the adult the vertebral column is composed of twenty-eight bones, which are divided into five distinct regions. The intervertebral discs are composed of fibro-cartilage, forming symphysis articulations between the vertebrae of the cervical, thoracic, and lumbar regions. The vertebral components of the sacral region are fused without discs and in the coccyx the vertebrae are reduced and attached by fibrous tissue rather than discs. During embryonic development, curves develop in the sacrum and thoracic regions and are considered primary curves. After birth, two additional curves develop; one in the cervical region as holding the head up develops, and one in the lumbar as sitting up occurs, followed by standing. These are considered to be secondary curves. The vertebral column is supported by muscles and ligaments attached to the vertebrae, permitting limited movement in most of the column. Intervertebral foramen develop between most of the vertebrae; these permit the spinal nerves to exit the spinal cord to innervate the body.

When describing vertebrae, the *body* (*soma, centrum*) of the vertebrae is the anterior portion of the vertebrae to which the intervertebral disc is attached. Attached to the body are vertical supports called the *pedicels*. The pedicel supports the *lamina*, forming the *vertebral foramen* housing the spinal cord. Lateral extensions of the lamina form the *transverse processes*; a vertical extension of the lamina forms the *spinous process*. *Superior articular surfaces* (*facet*) extend and articulate with the *inferior articular surface* (*facet*) of the vertebrae above it in the vertebral column. Modifications to this pattern of the vertebrae occur in the upper cervical vertebrae, sacrum, and coccyx. The thoracic vertebrae have additional articulations with the ribs, called *costal facets*, on both the body of the vertebrae and the transverse processes.

Cervical column

The cervical region consists of seven vertebrae. Each vertebra has foramenia in the transverse process named the *transverse foramen*. The typical cervical vertebra has a spinous process that is Y shaped at the tip. The first cervical vertebra (C1) lacks a distinct body, with reduced spinous and transverse processes. It is named the *atlas* and articulates with the occipital condyles of the skull, permitting a rocking motion forward and back. The second cervical vertebra has a modified extension of the body called the *dens* (*odontoid process*), which articulates with the atlas in the area where its body would be located and is held in place by a ligament. This permits a rotation of the head-neck side to side and is named the *axis*. As the cervical vertebrae come closer to the thoracic, they become more like those.

Thoracic column

The thoracic region consists of twelve vertebrae that articulate with the cervical vertebrae and each following thoracic vertebra until the lowest articulates with the lumbar vertebrae. Each thoracic vertebra articulates with a pair of ribs laterally, forming the thoracic cage. The cage moves up and down, changing volume during breathing as inhalation or exhalation occurs.

Lumbar column

The lumbar region of the vertebral column consists of five robust vertebrae with reduced but thickened transverse processes and spinous process. This region is supported by the interaction of the intervertebral discs, ligaments along the spine, and the muscles of the lower back and abdomen. The lumbar region tends to more mobile than the thoracic region.

Sacral column

The sacral region is composed of five elements, which are fused during embryonic development into the sacrum. The elements fuse at the body, transverse processes, and lamina. The fusion forms a series of sacral foramen on the anterior and posterior surface of the sacrum. The shape of the sacrum differs by gender. The male sacrum is typically curved anteriorly at the lower end, while the female sacrum is flattened, which is an adaptation for giving birth.

Coccyx column

The coccyx is typically composed of four highly reduced structures, consisting of the body with the elements being fused together.

Thoracic Cage

The thorax is the torso of the axial skeleton, commonly called the chest. It consists of thirty-seven bony elements that enclose the heart and lungs, providing protection for these vital structures. The bony elements are twelve thoracic vertebrae articulating with twelve pairs of ribs, with ten pairs of the ribs attaching to the elongated sternum by *costal* (hyaline) *cartilage*. The first seven pairs of ribs attach directly to the sternum and are identified as true ribs. Pairs eight, nine, and ten attach through the cartilage of the seventh pair to the sternum and are identified as false ribs. The last two pairs do not attach to the sternum and are identified as floating ribs. The first pair of ribs attach to the *manubrium* of the sternum. The next six pairs attach to the *body* (*gladiolus*) of the sternum. The inferior end of the sternum is the *xiphoid process*, which extends over the liver in the abdomen.

Appendicular Skeleton

The appendicular skeleton is composed of approximately 126 bones, which are freely movable. It is divided into an upper and lower grouping, attached to the axial skeleton by girdles. The upper girdle is the *pectoral girdle* and the lower is the *pelvic girdle*. The arms are attached to the upper girdle and the legs to the lower.

Upper Appendicular Division

Pectoral girdle

This girdle is composed of four bones forming the shoulders of the skeleton. The *clavicle* (collar bone) is attached to the manubrium of the sternum, then curves forward initially, then rearward, having an S shape. The lateral end articulates with the *scapula* at the *acromial process*, forming the superior portion of the shoulder. The acromial process is formed from the *spine of the scapula*. The scapula is attached to the spinous processes by muscle and ligaments on the medial margin and forms an oval depression laterally where the humerus articulates named the *glenoid fossa* (*cavity*). A process extends from the superior margin of the glenoid cavity anterior and slightly below the acromial process named the *coracoid process*. Ligaments attaching the clavicle and the acromial and coracoid processes form an encapsulation of the humerus, creating the freely movable, diarthrosis, *ball-and-socket* articulation of the shoulder, which allows motion in all three planes.

Upper appendages (arm)

The upper appendages are composed of sixty bones, divided into a right and left arm and hand. The *humerus* is the first bone, being the upper arm, which on the proximal end articulates with the scapula, forming the shoulder joint where the *head of the humerus* fits into the glenoid cavity of the scapula. The ligaments attach to the lateral surface of the humerus on the *greater tubercle* and the *lesser tubercle*, located on the medial anterior surface below the head. A groove is formed between the tubercle permitting nerves and blood vessels to move down the arm from the shoulder; this is called the *intertubercular groove (sulcus)*. On the lateral surface of the humerus, approximately one-third of the distance from the head, appears an elongated raised area where the deltoid muscle inserts, the *deltoid tuberosity*. Proceeding distally on the humerus on the medial side, it flares outward toward the axial skeleton where ligaments attach, forming the capsule of the elbow; this flared area is the *medial epicondyle*. A slight flaring occurs on the lateral side where the capsule attaches, called the *lateral epicondyle*. The humerus at its distal end articulates with the *radius* laterally and *ulna* medially. The radius articulates with the *lateral condyle (capitulum)* only seen on the anterior surface of the humerus, which allows the rotation of the radius across the ulna as the forearm pronates or supinates. The ulna articulation with the humerus is more complex in that the *medial condyle (trochlear)* is spiral shaped, permitting movement of the forearm of a range between 30 to 180+ degrees, which is the *hinge joint* of the elbow. The head of the ulna forms a *trochlear notch* with the *olecranon process (funny bone)* and *coronoid process*, which surround the medial condyle, forming the joint. Lateral on the margin of the head below the notch is a curved area, the radial notch, where the radius pivots against the ulna. On the posterior of the humerus is a fossa in which the olecranon process moves when the forearm is extended, and on the anterior surface is a fossa where the coronoid process extends when the forearm is flexed. The radius and ulna are attached to each other by an interosseous membrane. The radius has an insertion point from the biceps brachii muscle called the *radial tuberosity*. On the distal end of both the radius and ulna are styloid processes; the ulna's forms the bump on the small finger side, and the radius' on the thumb side at the wrist. The radius has a curved, medially located ulna notch, which permits the movement of the radius over the ulna. The wrist is formed by a series of eight small bones called *carpals*, which glide across each other during the movement of the wrist. They are held together by a ligamentous band attaching them to the radius. The palm of the hand is composed of five *metacarpals*, which articulate proximally with the carpals and distally with the *phalanges*. There are two phalanges that make up the pollux (thumb), and the remaining fingers are composed of three phalanges, making a total of fourteen in each hand.

Lower Appendicular Division

Pelvic girdle

The pelvis is composed of three bones; two of these bones are appendicular bones called the *os coxa*, and the other is the sacrum of the axial skeleton. The os coxa is formed by the fusion of three embryonic bony elements—the *ilium, ischium,* and *pubis*— fused into the single os coxa, that articulates with the sacrum, forming the pelvic girdle. The articulation of the os coxa with the sacrum is the sacroiliac articulation, a fibrous, slightly movable articulation on the posterior surface connecting the sacrum and the os coxa. Anteriorly, the pubic portions of the os coxa articulate with each other, forming the *pubic symphysis*, held together by a fibro-cartilage disc. The fusion of portions of each of the embryonic bony elements creates the socket *acetabulum*, in which the head of the femur articulates, creating the ball-and-socket joint of the lower appendages. The pelvis varies by shape for each gender as well as the individual's history. The major variation is the pubic angle, which is approximately 90 degrees for a male and 110 degrees or more for a female. Other variations occur with the iliac crest and pelvic tilt for post-gravid females.

Lower appendages (legs)

The lower appendages consist of sixty bones divided into a right and left leg. The femur proximally articulates with the girdle, with a lateral *greater trochanter* being the attachment point of the ligaments, and the *lesser trochanter*, located posterior medially, being the attachment location of tendons. Distally the femur articulates with the *patella (knee cap)* and the *tibia*. The patella glides across the anterior surface of the knee, providing protection. The tibia articulates with both the medial and lateral condyles of the femur, forming the hinge joint of the leg. At the head of the tibia, medially located on the anterior surface just below the knee, is the *tibial tuberosity*, where the *patellar ligament* inserts. Posteriorly and laterally the *fibula* articulates below the head of the tibia. The distal end of the tibia rests on the tarsal bone *talus*, creating the ankle. Medially the *medial malleolus* of the tibia stabilizes the articulation and laterally the *lateral malleolus* of the fibula stabilizes the articulation. There seven tarsal bones. The largest is the *calcaneus* (heel), followed by the *talus, navicular, cuboid,* and three *cuneiform* bones. There are five metatarsals and fourteen phalanges, with the same distribution as found in the hand. There are two phalanges in the *hallux* (great toe).

Articulations

Articulations are the contact areas of two or more bones. The articulations are grouped by mobility of the articulations and by the types of attachment of the bones to each other. The description that will be applied here is the mobility, with secondary descriptions of the attachments.

Synarthrosis articulations are immovable. These articulations are rigid, with no synovial cavity or membrane. These articulations are attached through fibrous connective tissue or by hyaline cartilage. *Sutures* are immovable articulations found in the skull and the flat bones, and are attached by fibrous connective tissue to form a very tight junction that tends to ossify over time. *Gomphosis* is an articulation formed by a socket (depression) in one bone permitting a conical shaped portion of the other bone to be inserted, attached together by a band-shaped fibrous tissue. An example is the teeth in both the maxilla and mandible, which are attached by the periodontal ligament. *Synchondrosis* articulations are bony elements attached together by hyaline cartilage, such as that found in the epiphyseal plate of the long bones.

Amphiarthrosis articulations are slightly movable. These articulations lack a synovial cavity and are attached by either fibrous tissues or fibro-cartilage. *Syndesmosis articulations* are found between the radius and ulna, and tibia and fibula, and are attached by an *interosseous membrane*, a fibrous membrane, and the fibrous bands associated with the ankle and wrist. *Symphysis articulations* attach the bones together by fibro-cartilage discs. Examples of symphysis articulations are found in the pelvis, with the pubic symphysis, and the intervertebral discs, which permit a slight movement.

Diarthrosis articulations are those which are freely movable. They are referred to as synovial articulations because they have a synovial cavity, a joint capsule composed of fibrous connective tissue (ligaments) and the ends of many of the bones covered with hyaline cartilage; the articulation is lubricated by synovial fluid. The movement of these articulations creates movement based on the development of leverage and tension when associated with muscles bridging the articulations.

Gliding articulations are constructed by the articulation of flat surfaces moving back and forward or side to side in one plane. They are described as uniaxial, with the movement limited by associated ligaments. These articulations are found between the carpals in the hand and the tarsal bones of the feet.

Hinge articulations are constructed of two bones, one having a concave surface riding over the convex surface of the other. This permits movement by increasing and decreasing the angle for the articulating bones, which is defined as *flexion* and *extension movement*. Since this movement occurs in a single plane, this articulation is uniaxial. The movement between the femur and tibia at the knee and the ulna and humerus at the elbow are hinge-articulations.

Pivot articulations are constructed of a raised surface in one bone fitting into a ring-like depression in another, permitting rotation around a fixed point. In some cases they are attached by ligament. An example is the radius with the lateral condyle of the humerus, and the radius and ulna. Some suggest that the articulation between the first and second cervical vertebrae is a pivot type, but some characteristics are absent. This type of articulation is considered to be uniaxial.

Ellipsoid (condyloid) articulations are constructed of an oval condyle on one bone articulation in an elliptical cavity in another bone, permitting movement in two planes, defining it as biaxial. Movement allowed is flexion, extension, protraction, retraction, abduction, and adduction, as well as circumduction. These movements are encountered in the radius articulating with the carpals, forming the wrist, and the mandible articulating with the temporal bone, forming the jaw allowing chewing motion.

Saddle (sellaris) articulations are composed of one bone having a saddle-like shape, with the other bone riding astride. This is found in the articulation between metacarpal I and the trapezium (carpal), permitting motion of flexion, extension, abduction, adduction, and circumduction of the thumb. This is a biaxial articulation. Some suggest that the ankle is a saddle articulation composed of the tibia resting on the talus, stabilized by the fibula laterally, since the ankle is capable of the biaxial movement.

Ball-and-socket articulation is constructed of a round, ball-shaped head on one bone with a curved depression of another bone of varying depth. This articulation permits movement in three planes, and so is defined as triaxial. Examples of the ball-and-socket articulations are found between the scapula and humerus, as well as between the os coxa of the pelvis with the femur.

These freely movable articulations permit movement over a wide range. Most movement results from the interaction of several articulations. Examples of movements are *elevation-depression*, which can consist of the head, shoulder, or entire body indicating a person's psychological status under existing conditions; *protraction-retraction,* which can consist of the mandible, shoulder, or entire body and indicate aggressive or submissive body language. The pointing of toes is *dorsiflexion* or *plantar flexion* (this happens with a person wearing heels), and involves the entire foot. Movement of the ankle rotating the sole of the foot medially or laterally (think of a baby playing with their feet between three and six months) is called *inversion* or *eversion. Supination-pronation* is the can of the anatomical position from one position to the other. It may involve the entire body being placed face up or face down or the forearm rotating the hand palm up or palm down.

LABELING ACTIVITY

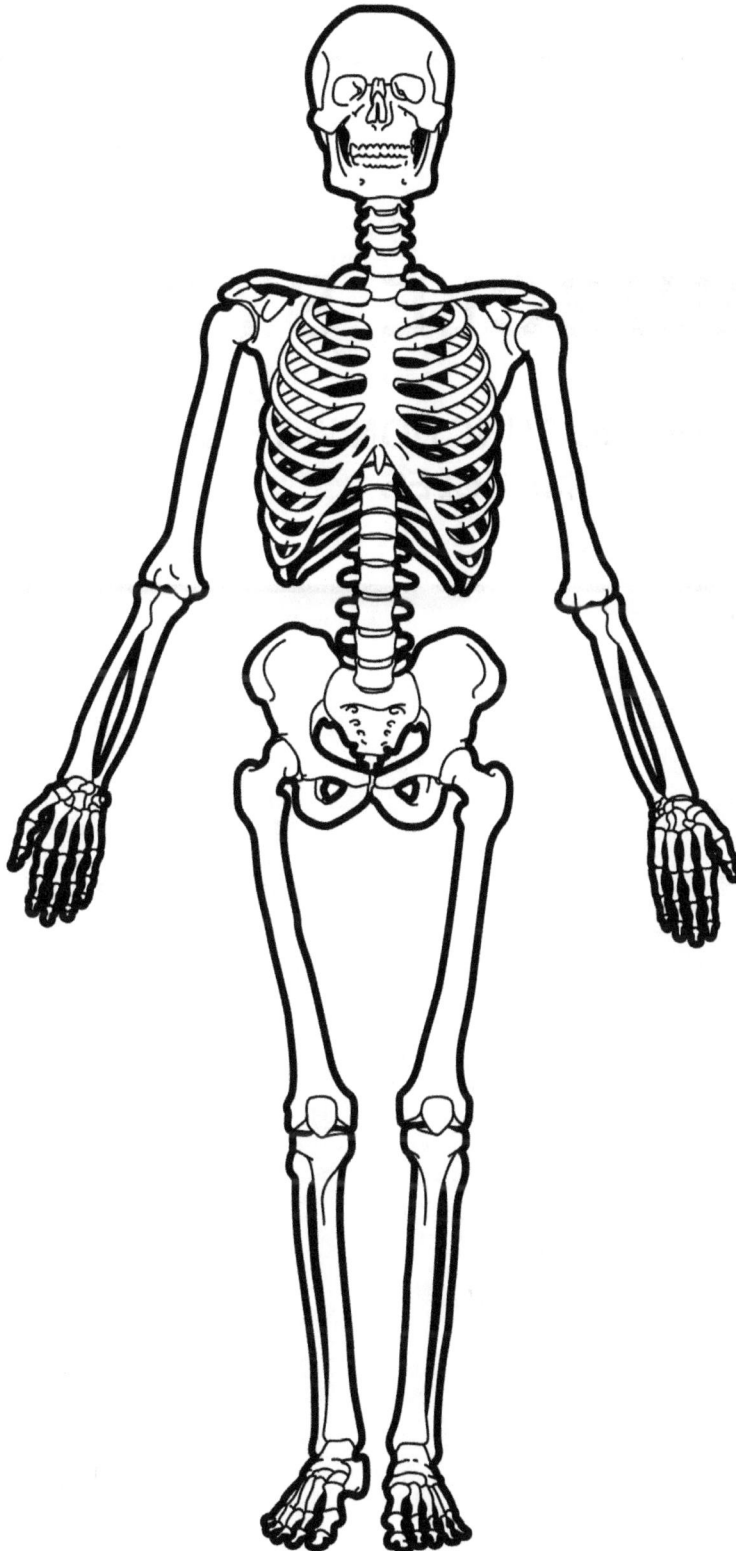

Fig. 5.1 Source: Copyright © 2013 Depositphotos/Chisnikov.

CHAPTER SIX

THE MUSCLE SYSTEM

ORGANIZATION AND FUNCTION WITH EMPHASIS ON SKELETAL MUSCLES

Upon completion of the chapter readers will have the essential understanding and knowledge of muscle anatomy and physiology:

1. Description and comparison of the types of muscle tissues
2. The specific organization of anatomy and physiology
3. The movement generated by skeletal muscle in relationship to the skeleton

The muscle system creates motion throughout the body by shortening the fibers, creating tension, followed by relaxation; this provides leverage that permits lifting and maintenance of posture. Muscle tissue provides a supporting framework in the body and varying the shape of organs. The muscle tissue by its metabolism during contraction creates heat, which aids in maintaining body temperature.

Muscle Tissue Characteristics

All muscle tissues respond to stimulation, which makes the cells *excitable* (*irritable*), with the stimulus creating a change in the cells' resting *action potential*. When stimulated, the cells respond by shortening and thickening, generating tension on the surrounding connective tissue applying force to some location; the cells are *contractile*. The resting muscle cells can be stretched with the movement of other body parts without damage, which makes the cells *extensible*. Muscle cells return to the resting shape following contraction or stretching, which makes them *elastic*. These characteristics are present in all types of muscle tissue; how each tissue varies with each characteristic is dependent on the location and function of that muscle or muscle tissue.

Types of Muscle

There are three types of muscle, differing in their internal organization, location, and performance. The *skeletal* (*striated*) *muscle* is muscle tissue attached directly or indirectly to the skeletal system. The cells merge, creating multi-nucleated myofibrils that are elongated and internally organized into *sarcomeres*, giving the appearance of striations. These myofibrils are individually wrapped in a layer of connective tissue requiring that each be stimulated by nerves and contraction is under conscious control, making these muscles voluntary. The majority of this chapter will be devoted to the physiology and description of skeletal muscle. Skeletal muscle generates its energy requirement by three methods: creatine phosphate which is produced in the resting muscle; glycolysis of carbohydrates which result in the formations of lactic acid, and aerobic respiration when adequate oxygen is available from various sources.

The *cardiac muscle* composes the wall of the heart and is referred to as the myocardium. These cells are organized internally into sarcomeres, with cells being branched and in contact with at least four other muscle cells through a tight gap junction called the *intercalated disc*. The cardiac muscle cells have a single, centrally located nucleus and contraction is internally stimulated in the heart, so this type of tissue is involuntary. Cardiac muscle obtains it energy for contraction by aerobic respiration only.

The *smooth (visceral) muscle* is located in internal organs and dermis. The cells are spindle shaped with a centrally located nucleus. The contracting elements (*actin* and *myosin*) are organized in a net-like arrangement beneath the cell membrane, permitting elongation of the cell. The smooth muscle tissues are involuntary and can be stimulated by a variety of mechanisms. Smooth muscle tissues are arranged in layers with all the cells oriented in the same pattern. Energy for contraction is derived by a unique path way identified as the myosin light chain reaction to produce ATP. Troponin which is a regulatory protein in muscle is replaced by *calmodulin* in smooth muscle.

Skeletal Muscle Description and Physiology

Skeletal muscle cells must each be stimulated by a nerve impulse to initiate contraction. The nerves that stimulate the cells are *motor nerves*. They contact the muscle cell at an *axon terminal*. Normally a single motor nerves stimulate multiple myofibrils, not just one. The motor nerve and all of the myofibrils it contacts (innervate) compose a *motor unit*. The neuron passes the action potential along its axon to each of the myofibrils, initiating the change in the myofibrils' resting action potential, which triggers a series of external and internal events. The number of myofibrils controlled by a motor nerve ranges from 10 to 2,000, with an average of about 150. The location where the contact is made between the motor nerve axon terminal and the myofibril is the *neuro-muscular junction*. The neuro-muscular junction consists of the *axon terminal bulb*, a space called the *synapse*, and the *motor end plate* on the myofibril.

Physiology of Stimulation, Contraction and Relaxation

The arrival of the action potential (impulse) at the axon terminal bulb initiates a series of events that results in the contraction of the myofibril. The action potential initiates *calcium ions* to enter the axon terminal from the synapse through voltage-gated calcium channels. As this occurs, the axon terminal bulb releases vesicles containing the neurotransmitter *acetylcholine* into the synapse. The acetylcholine migrates to the motor end plate on the myofibril and binds with acetylcholine receptors, triggering action potential change on the *sarcolemma* (plasma membrane) as *potassium ions* move out of the cell and *sodium ions* move into the cell. This results in a voltage shift on the interior of the myofibril from –70 millivolts to a + 30 millivolts called *depolarization*. This voltage shift spreads over the surface of the myofibril and into the interior through the *T (transverse) tubules*. In the interior of myofibril, the T tubules are closely connected to the *sarcoplasmic (endoplasmic) reticulum*; the voltage shift initiates voltage-gated channels to release calcium from the sarcoplasmic reticulum into the cytoplasm. The calcium diffuses into the sarcomeres, where the calcium binds with *troponin*, which is a regulatory protein that blocks the myosin binding sites. As calcium troponin complex forms, the myosin binding sites are exposed by the rotation of *tropomyosin* (regulatory protein that physically blocks the myosin binding sites). Cross bridges form between the actin and myosin strands and the actin strand is pulled to the center of the sarcomere, generating tension of the myofibril by shortening the myofibril. In order for the cross bridge formation to occur, the myosin heads form the cross bridge and must be cocked, which requires ATP. On the *myosin head* is a binding site for ATP, which binds then releases energy to cock the myosin head, but retains the ADP and the phosphate ion. Once the head binds with actin, the phosphate ion is released, with the ADP retained in the binding site. Once the actin strands are pulled to the center of the sarcomere during the *power stroke*, the ADP is released and is replaced by a new ATP, breaking the cross bridge as it releases its high energy bonds, cocking the myosin head.

The action of the muscle is divided into three phases: the initiation phase occurs from the nerve impulse arriving at the axon terminal to the calcium troponin complex being formed; the contraction phase exists from the tropomyosin rotating from the myosin binding sites until the release and cocking of the myosin heads; and the third phase is the recovery and relaxation of the myofibril. The recovery phase begins with the release of the myosin heads from the actin and the lengthening of the sarcomere, returning it to the resting state. Relaxation of the myofibril begins when no additional impulses reach the axon terminal so that no further acetylcholine is released, and that which is present in the synapse and that released from the motor end plate receptors is broken down into acetyl and choline units by *acetylcholinesterase*. Ion pumps in the sarcolemma are activated by ATP attachment to the integral protein channels, moving sodium out of the myofibril and potassium into the myofibril against the concentration gradient. Internally, calcium ion pumps driven by ATP move calcium into the sarcoplasmic reticulum, where it is stored by *calsequestrin*. Once the myofibril returns to it resting membrane potential it has achieved relaxation.

Skeletal Muscle Description

Skeletal muscle is composed of myofibrils because individual muscle cells merge together, forming the structures present in the normal skeletal muscle. Each myofibril is covered by a thin layer of connective tissue called the *endomysium*, which merges with others to form *fasciculi*, and is covered by another layer of thin connective tissue (the *perimysium*), creating organized bundles of contracting tissue that are connected to the skeleton in various configurations. These are covered by a third layer of connective tissue, *epimysium*, which merges to form the *tendon*, which attaches the muscle to the bone. The arrangement of the fasciculi with the tendon and the attachment of the muscle to the skeleton will define the leverage and force capability of the individual. The shape of skeletal muscle is defined by its fascicular arrangement. The fascicular patterns consist of the myofibrils oriented parallel to the long axis of the body, which is gives rise to the name *rectus*, such as the rectus abdominis. The fascicular arrangement the myofibrils that runs parallel to the long axis, tapering toward the tendons, is a *fusiform* configuration, which is represented by the biceps brachii muscle. Another arrangement is a *triangular* pattern of the fasciculi, which merges to an apex such as that encountered in the pectoralis major muscle. The myofibrils arranged in a *circular* pattern are generally named *orbicularis* muscles. Fasciculi arranged with a central attachment at an angle are identified as pinnate, only if arranged like a feather, *bipinnate*, like a leaf or *multipinnate*.

Skeletal muscle has limited power to regenerate after the age of one. As we grow, skeletal muscles increase in size by enlargement of existing cells as they respond to stress applied by exercise. The increases are due to additional mitochondria added to the muscle by conditioning. Conditioning of the skeletal muscles stimulates an increase in the blood supply and the efficiency of electrolyte movement. Most skeletal muscles are a mix of myofibrils: those that store large amounts of oxygen in *myoglobin* are called *red muscle*; those that have little myoglobin are called *white muscle*. The difference in the red and white fibers is the response to stimulation without fatigue. All the myofibrils within a motor unit are the same—either all red or all white; motor units are not mixed. Each named muscle is stimulated by a number of motor units. Skeletal muscle is described by the pattern of contraction and endurance of the muscle. The myofibrils that produce their energy by an aerobic mechanism are identified as *slow twitch slow type I oxidative muscle* tissue and tend to be fatigue resistant. They tend to be the muscles involved in the maintenance of posture. *Fast twitch muscle* tends to be fatigue resistant and is identified as *fast oxidative type IIA*, which produces energy by very efficient aerobic means and a rich abundant blood supply. A third type of muscle meets its energy demand by glycolytic mechanisms; it is fatigable and is identified as fatigable, *type IIB*, or *fast twitch B muscle*. ATP in this type of muscle will be produced by a variety of mechanisms and, due to lack of efficiency in oxygen replenishment and waste removal. These muscles builds up lactic acids, resulting in fatigue.

The mechanisms used to produce ATP could be any of the following, or a combination based on conditioning and the demand placed on the muscle. *Creatine phosphate (phosphagen) system* uses creatine phosphate stored in the resting muscle as an immediate source of high-energy phosphate to produce ATP by substrate phosphorylation. The muscle produces creatine phosphate which is stored in the myofibrils for later use when initial peak demand for action occurs. Typically, muscles only are able to store enough creatine phosphate to support maximal muscle effort for a period of approximately fifteen seconds or the time required to cover one hundred meters. Glycogen is stored in skeletal muscle as a carbohydrate that is converted to glucose, which undergoes *glycolysis* to produce the ATP to support contraction. This process does not require oxygen and, if the level of oxygen is inadequate, produces lactic acid. Many muscle cells are capable of producing adequate ATP for approximately thirty to forty seconds when an initial demand for action occurs. This would be adequate for a distance of about 400 meters. The final mechanism used is *aerobic respiration*, or the complete breakdown of glucose to produce ATP by both glycolysis and electron transport following the tricarboxylic acid cycle. This requires adequate available oxygen and effective waste removal.

There are more than 700 skeletal muscles that have been named. Most are named by using some characteristics of the muscle. The *direction* of the muscle and its fibers could result in a muscle named transverse or oblique. The *location* of the muscle is another common characteristic and could result in a muscle named frontalis. Many muscles are named by *size*; for example, using *maximus* or *minimus* could be applied. Other muscles are named by the *origin*, the *insertion*, or the *action* of the muscle. Many times, the names are descriptive, using multiple characteristics for the name of the muscle. The most effective way to identify skeletal muscles is to locate them on your own body, especially the superficial muscles. Deeper muscles require the removal of the superficial muscles to locate them, which exceeds an introductory course.

CHAPTER SEVEN

THE NERVOUS SYSTEM

ORGANIZATION AND FUNCTION OF THE CENTRAL AND PERIPHERAL DIVISIONS AND SENSORY ELEMENTS

LEARNING OBJECTIVES

Upon completion of the chapter readers will have the essential understanding and knowledge of the nervous system and its sensory components:

1. The organization of the nervous system
2. The functions and locations of neurons and glial cells
3. Details of the central nervous system
4. Sub divisions of the brain and their functions
5. The spinal cord, its divisions, plexuses, and function
6. Details and components of the peripheral nervous system
7. Sensory components' distribution with emphasis on smell, taste, touch, vison, and hearing/equilibrium

The nervous system is composed of two major divisions: the central and the peripheral. The peripheral division includes the sensory components of the system. The central division is involved in the interpretation of the senses, regulating responses throughout the body, and sustaining communication. The primary cells of the nervous system are the *neurons*, which are the cells capable of initiating an action potential (impulse) and conducting the impulse it to another location. Neurons have a central body containing a nucleus, called the *soma*, in which the cytoplasm is thickened and called the *neuroplasm*. Impulses are received by the neuron from its *dendrites*, which conduct the action potential to the cell body. The impulse is passed through the cell body to the *axon*, which conducts the action potential to some other location. The impulse travels by localized membrane charge changes in a wave pattern and involves the movement of electrolytes into and out of the dendrites, soma membrane, or axon. There are several different neurons found throughout the body. One type of neuron is surrounded by dendritic like processes and is called *associative* or *anaxonic neurons*, which are in contact with multiple other neurons or cells. Another neuron has only axons attached to the soma. These are called *bipolar neurons* and are frequently located in the peripheral and spinal cord area. The soma is offset with two axons; this is identified as a *unipolar neuron*. The neurons that appear as typical, having the dendrites close to the soma and a clearly defined axon, are called *multipolar neurons*.

Neurons are incapable of caring for themselves. The support of the neurons is carried out by *neuroglia*. The glia protect and support the neurons and keep the impulses contained so that they are conducted as required without degradation. Some act as phagocytes in the central system, some produce the cerebrospinal fluid, while others isolate the signals from one another. Those in the central system are the *astrocytes, microglia, ependymal cells*, and *oligodendrocytes*. The ones in the peripheral system are the *Schwann cells* and the *satellite cells*. Neuroglia cells have the capacity to divide, while the neurons lack that capacity. Under specific conditions portions of peripheral neurons may be repaired or regenerate, but never the entire neuron.

Central Nervous System

The central nervous system consists of the spinal cord and the brain, encapsulated in bony elements for protection. Beneath these bony elements are the *meninges*. The outermost membrane is a tough, fibrous *dura mater* that lines the bony cavities. The middle layer is the *arachnoid mater*, which is thin with spider web-like collagen and elastic fibers throughout. Beneath the arachnoid is the *subarachnoid space*, which is filled with clear *cerebrospinal fluid*. Covering the surface of the brain and spinal cord is the thin, transparent *pia mater*.

Brain

The brain consists of the cerebrum, diencephalon, midbrain, pons, medulla oblongata, and cerebellum. Each of these subdivisions of the brain has a unique function that occurs in coordination with all of the other subdivisions.

Cerebrum

The cerebrum is the region of the brain responsible for thought, intellect, memory, muscle regulation, and sensory analysis. It is divided by the *longitudinal fissure* into a left and right hemisphere. The hemispheres are interconnected by the *corpus callosum*, which appears as a thick band of white matter. The outer surface of the cerebrum is gray matter, which consists of non-myelinated neurons beneath is a layer of myelinated connections forming the white matter. Each hemisphere is divided into lobes that are interconnected by anterior commissures, projection fibers, and longitudinal fascicules. The *frontal lobe* has been found to control behavior and voluntary muscle activity. The *parietal lobe* functions to locate and evaluate touch, pressure, pain, vibration, taste, and temperature. The *occipital lobe* recognizes and evaluates visual stimuli. The *temporal lobe* evaluates auditory (sound) and olfactory (smell) stimuli. The *insula region* is a small area that pays a role in language understanding, taste, and integration of sensory input from viscera. Beneath the cerebrum extend lateral ventricles, which permit the circulation of cerebrospinal fluid bathing the neurons in nutrients and removing wastes.

Diencephalon

The diencephalon surrounds the *third ventricle* where the cerebrospinal is formed from the blood by the choroid plexuses. It is inferior to the corpus callosum and contains three major functional regions: the thalamus, epithalamus, and hypothalamus. The *thalamus* routes sensory impulses to appropriate locations in the cerebrum, and motor impulses along appropriate routes. The *epithalamus* is the most superior portion, containing the *choroid plexuses* where the blood is filtered and cerebrospinal fluid is formed; it is the site of the blood-brain barrier. It contains the *pineal gland*, which produces melatonin that regulates our biological clock (sleep–wake cycles). The hypothalamus is divided into six functional areas governing different functions of the body. The *mammillary bodies* control the reflex that results in licking and swallowing vital to newborns learning to nurse. The *tuberal nuclei* controls the release of hormones that in turn control the actions of the endocrine cells of the adrenal gland. The *supraoptic nuclei* secrete antidiuretic hormone (ADH) to restrict water loss by promoting water retention in the kidneys. The *paraventricular nucleus* secretes oxytocin to strengthen smooth muscle contraction, particularly during labor. The *preoptic area* regulates the body's temperature and is sensitive to stress. The *suprachiasmatic nucleus* coordinates our day–night activities and is influenced by the surge of sex hormones during puberty.

Midbrain

The midbrain extends from the pons to the diencephalon and contains a cerebral aqueduct that circulates cerebrospinal fluid between the third and fourth ventricles. It is composed of tracts conducting motor impulses to the spinal cord, medulla, and pons. Reflex centers for visual activities, such as tracking and scanning, and startle reflex, are located in this area.

Pons

The pons relays impulses involving cranial nerves V, VI, VII, and VIII. It also adjusts the pattern of breathing (respiratory) through two centers: the *apneustic* center, which controls the depth of breathing, and the *pneumotaxic* center, which controls the rate of breathing. Both are located in the medulla oblongata. It relays sensory impulses to the thalamus, motor impulses to the spinal cord, and sensory data to the cerebrum.

Medulla oblongata

The medulla oblongata (medulla) regulates the heart rate and force of contraction through the cardiac center. It regulates blood pressure and blood flow through the vasomotor center and, in conjunction with the pons, regulates respiratory functions. Sensory impulses are relayed to the thalamus and the cerebellum. A fourth ventricle lies between the medulla and the cerebellum, producing cerebrospinal fluid in a choroid plexus.

Cerebellum

The cerebellum has been referred to as "the little brain." It functions to evaluate and coordinate activities of skeletal muscle to maintain balance and coordinate movement. It directs impulses from other areas to maintain posture and equilibrium. This is one of the areas of the brain most effected by medications and beverages.

Spinal Cord

The spinal cord is continuous with the medulla of the brain stem. It has a central canal that receives cerebrospinal fluid from the fourth ventricle's choroid plexus. This cerebrospinal fluid passes down the spinal cord and empties into the *epidural space* around the level of the lumbar vertebrae. Like the brain, the spinal cord is divided into hemispheres by a posterior median sulcus and an anterior median fissure. Unlike the brain, the white matter is external to the gray matter. The spinal cord is covered by extensions of the meninges, which extend to the sacral region and are filled with cerebrospinal fluid. The configuration of the spinal cord is consistent from the medulla to level of the first or second lumbar vertebra, where it changes into a series of nerve tracts. The spinal cord in the cervical region has an enlargement that provides innervation to the shoulder girdle and upper limbs. A second enlargement occurs in the lumbar area to provide innervation to the pelvis and lower limbs. The spinal cord after the lumbar enlargement tapers rapidly into the *conus medullaris,* from which extend a series of nerve tracts called the *cauda equine,* it visually resembles a horse tail. A further extension connects the spinal cord to the first coccyx vertebra to maintain tension of the spinal cord. It is the *filum terminale.*

From the spinal cord arise a series of nerves that develop into the peripheral nerve connections. There are thirty-one pairs that arise from the dorsal (sensory) root ganglion where the neurons bodies are located. The ventral (motor) roots do not have ganglia, since they are axonal tracts. Once the spinal nerves emerge from the intervertebral foramen, they are considered to be mixed, having both sensory and motor functions forming the peripheral system. The spinal cord is divided into regions similar to the vertebral column. The cervical region produces eight pairs of spinal nerves, they emerge to form two plexuses: the cervical plexus, composed of the first four cervical spinal nerves and the dorsal root of the fifth; and the brachial plexus, composed primarily of the ventral root of the fifth cervical through the first thoracic nerve, with an interconnection with the fourth cervical nerve. The *phrenic nerve,* which controls the diaphragm and muscles of breathing, has contributing pathways from the third, fourth, and fifth cervical nerves; damage to this nerve results in breathing difficulties. There are twelve pairs of spinal nerves emerging from the thoracic region. The lumbar region has five pairs of spinal nerves, with the first four pairs forming the lumbar plexus. Five pairs of spinal nerves emerge from the sacral regions, with the fifth lumbar and the first three sacral forming the sacral plexus. There is a contribution to this plexus by the fourth lumbar nerve. The coccygeal nerve and the last two sacral nerves form a plexus that innervates a small area of the pelvis.

Many of the reflexes that are immediate responses to a stimulus occur in the spinal cord as a reflex arc. A reflex arc is composed of a sensory nerve, an interneuron that connects to both the sensory and motor neurons, and the motor neuron. When stimulated, the sensory impulse reaches the spinal cord, the impulse splits a portion passes to the brain, and another portion of the impulse stimulates the interneuron. The interneuron stimulates the motor neuron sooner than a motor impulse can be sent from the brain, resulting in movement to respond to the original stimulus.

Nerve conduction varies with the size of the nerve fiber, the temperature of the nerve, and how the fiber is or is not protected by the insulating myelin sheath. A warm nerve fiber resulting from a fever will conduct impulses faster and more frequently than at normal body temperature. The nerve fiber will conduct an impulse faster when myelinated than when not. Since the myelin sheath is not complete along a nerve fiber, it has a series of sheathed areas and areas where the sheath ends create nodes where the impulse slows and then speeds up again in the next sheathed area.

Peripheral Nervous System

The peripheral nervous system consists of all nerve activity outside of the brain or spinal cord. It is composed of sensory and motor functions. The sensory function ranges from the detection of light touch, chemical relating to smell and taste, light, and waves of pressure in air which are hearing and vibration. The motor function ranges from the control of skeletal muscle to stimulation of digestive processes. The axon of these nerves are protected by myelinations produced by Schwann cells and are grouped together by a series of connective structures protecting the nerve bundles. It consists of the somatic division and the autonomic division.

Somatic Nervous System

The somatic division consists of sensory and motor functions at our level of awareness. These sensory functions include the special senses (vision, hearing, smell, taste, and equilibrium); somatic senses (pain, thermal, and touch); and proprioceptive senses (position of the body and its parts) in relationship to the surrounding external environment. The motor functions involves the stimulation of skeletal muscles to maintain position or movement in response to sensory stimuli and the maintenance of muscle tone.

Autonomic Nervous System

The autonomic division consists of sensory and motor functions that are involved with management of the body's internal environment, below our level of awareness. Centers in the hypothalamus and the brain stem are involved in the maintenance of the internal environment. There are three subdivisions of the autonomic system: the *sympathetic division* stimulates the internal organs and systems to prepare for and respond to stress, making energy available; the *parasympathetic division* stimulates internal organs to reverse course and return to a resting or relaxed mode, storing energy for the future; and the *enteric division* manages the processes of digestion.

Senses

The senses are the part of the somatic nervous system frequently referred to as special senses. Senses provide information about our external environment at the level of our awareness (conscious level). Receptors capable of responding to environmental stimuli, a nerve pathway to conduct the information as an impulse to the central nervous system, and a center in the central nervous system to interpret the impulse are essential for the senses to function.

Receptors respond to various stimuli from a variety of sources to determine the status of the environments and position within the environments. Some receptors respond to mechanical stimuli (*mechanoreceptors*), while others respond to changes in temperature (*thermoreceptors*). Damage to tissue stimulates changes in the tissues involved, causing *nociceptors* to respond the stimulus as pain, while others respond to light, called *photoreceptors*. Those responding to chemicals are *chemoreceptors* and others responding to changes in osmotic pressure are *osmoreceptors*.

Widespread over the surface of the body are receptors that detect touch, vibration, itch, and tickle. These are referred to as *tactile* sensations. These receptors are mechanoreceptors composed of encapsulated structures or free nerve endings. *Meissner's corpuscles* detect light and fine touch, and are located in the dermal papilla in the hands, eyelids, tongue, and other locations. Located deep in the dermis are the *Ruffini corpuscles*, which detect stretch, and the *Pacinian corpuscles*, which detect pressure and vibration. In the base or papilla of the hair follicle is a plexus of free nerve endings that respond to the movement of the hair shaft. These receptors respond to the external stressors such as gentle pressure, heavy pressure, vibration, and other physical influences on the surface of the skin.

In the skin there are other receptors that detect the external thermal environment. Located in the *stratum basale* are cold receptors, which are capable of sensing temperatures between 50 and 105 degrees Fahrenheit (10 to 40 degrees Celsius). In the dermis are warm receptors, which sense temperature ranges between 90 and 118 degrees Fahrenheit (32 to 48 degrees Celsius). Temperatures above these ranges are considered to be hot and below these ranges are considered to be cold. The most sensitive location to temperature is the anterior surface of the skin at the wrist.

Specialized senses

The sense of *smell* requires that the material to be detected must be water- or lipid-soluble and volatile (gaseous or forming microdroplets) to contact the chemical receptors (chemoreceptors) located in the superior of the nasal mucosa. The olfactory mucosa consists of between 10 to 100×10^6 *olfactory* receptors, connected to the olfactory bulb through the olfactory tract to the hypothalamus and limbic area, to the temporal lobe where interpretation of the smell occurs. Our olfactory senses are very sensitive and capable of detecting many odors; they do have a problem if the receptors become rapidly saturated, losing the perception of the odor.

The sense of *taste* requires that material to be detected be water- or lipid-soluble to stimulate the chemo receptors located on the tongue and oropharynx. Taste consists of sweet, sour, bitter, salty, and umami (savory taste and texture of meat or protein). Over the surface of the tongue, soft palate, and oropharynx, it is estimated that there are 10,000 taste buds, distributed with bitter receptors toward the tip of the tongue and sour receptors toward the oropharynx. Approximately two-thirds of the taste buds on the anterior tongue are innervated by the *facial nerve* and those on the posterior third are innervated by the *glossopharyngeal nerve*. Taste buds located in the oropharynx are innervated by the *vagus nerve*. Impulses are routed from the medulla oblongata to the hypothalamus, limbic system, thalamus and then to the parietal lobe for interpretation.

The sense of *sight* (vision) contributes more than half of all sensory receptors and the largest portion of the cerebral cortex devoted to impulse interpretation. The eye is the principal structure that responds to light stimulation, although melanocytes in the skin respond as well.

Surrounding and supporting the eye are a series of structures that protect the eye from dust, debris, direct sunlight, and sweat, such as the eyebrows and eyelashes. The eyelid and the *lacrimal glands* wash, moisten, and clean the surface of the eye. The tears produced by the lacrimal glands are composed of water, electrolytes, and a series of bacteriostatic lysozymes that are spread across the eye surface by the eyelids. Groups of six extrinsic muscles move the eye to center visual objects. The *conjunctiva*, a mucus membrane, protects the posterior of the eye by sealing the socket and being a site that stimulates immune reactions. The eyeball is approximately one inch in diameter, with only one-sixth being exposed beyond the conjunctiva. It is composed of three distinct layers called tunics. The outermost layer, the *fibrous tunic*, is divided into the clear transparent *cornea* and the dense, fibrous, white *sclera*, which provides protection of the inner structures and the cornea permits the entry of light into the interior. The middle layer, the *vascular tunic*, consists of the *choroid layer*, which lines the posterior portion of the eye; being highly vascular, it provides nutrients to the retina and removes waste. The *ciliary body*, consisting of the ciliary processes, suspensory ligaments, and ciliary muscles, is attached to the lens, changing its shape to accommodate the focusing of an image on the retina. The last portion of the middle layer is the *iris* and *pupil*, which control the amount of light that contacts the lens and transmits it to the retina. The innermost portion is the *retina*, which lines the posterior compartment with an optic disc at the entry of the optic nerve. It consists of a pigmented layer containing melanin to control light scatter, and a neural layer composed of rods, cones, and supporting neurons. The *cones* respond to intense light and are of three types, responding to various portions of the light spectrum, permitting color vision. The *rods* respond to lower light intensity and enable our vision at night. The most intense concentration of cones is in the *macula lutea* around the *fovea centralis*. Moving around the posterior compartment, the further from the macula the more rods and fewer cones are found. Internally the eye is divided into two anterior chambers and a posterior compartment. Behind the cornea is the anterior chamber, which is filled with a watery *aqueous humor* that extends to the iris. Behind the iris, extending to the lens, is the posterior chamber, which is filled with the aqueous humor. The posterior compartment, containing the retina, is filled with a jelly-like *vitreous humor*, which provides gentle support to the retina. If the vitreous humor liquefies, it loses the ability to support the retina, resulting in tears to the retina.

Vision is the reflection of light from an object that is detected by the eye. The light is collected by the cornea, which bends the light to pass through the liquid aqueous humor and pupil to the lens. The lens flattens or becomes more curved to focus the light on the retina. The crisp, focused image stimulates the cones and rods, which create an impulse that is conducted by the *optic nerve* to the *optic chiasma*, located below the hypothalamus, where the impulse from each eye is split into unequal portions and routed to the opposite portion of the occipital lobe, with a portion of the impulse routed to the same occipital lobe, providing binocular vision. If the lens focuses the image in front of the retina, a person is *myopic* (nearsighted). If it focuses behind the retina, a person would be *hyperopic* (farsighted). Astigmatism is when the eyeball shape is not spherical, which rotates the image slightly and is corrected by adding a rotation into the lens. Color vision is based on the action of the three types of cones in the retina which respond to different portions of the light spectrum providing the range of color we see. Red-green abnormalities are the most common in males and rare in females. True color blindness would require that an individual have no functioning cones, and only rods.

The ear is the primary structure involved in both hearing and equilibrium. Both respond to stimuli of a mechanical nature. Hearing responds to the wave motion and energy levels in the air, while equilibrium responds to the energy involved with changes involving the force of gravity or change in velocity, which translate into static and dynamic equilibrium.

The anatomy of the ear can be divided into three regions. The outer region, consisting of the *pinna* (auricle), collects the waves in the air and directs them into the external auditory (acoustic) canal. These waves travel down the canal until contacting the *tympanum* (eardrum), causing it to flex. In the middle ear, connected to the tympanum, are three bony ossicles: the *malleus* is connected to the tympanum and is vibrated as the tympanum is moved by the pressure waves; the *incus* is attached to the malleus loosely and is moved by the movement of the malleus; and the third is the *stapes*, which is loosely connected to the incus and is moved when the incus vibrates. The ossicles are held in place in the middle ear by small muscles. The middle ear is normally filled with air and is vented by the *Eustachian tube*, connected to the throat; this equalizes the pressure on the tympanum, allowing free movement with the sound waves. The inner ear is connected to the stapes at the oval window. The motion of the stapes creates waves in the *perilymph* of the *scala vestibuli*. This wave motion creates motion in the basement and vestibular membranes, translating into movement of stereocilia in the inner hair cells of the *organ of Corti* in the *scala media*. This movement triggers ion changes, which generate the impulses we recognize as sound. These impulses are conducted by the *cochlea branch* of the *vestibulocochlear nerve* to the thalamus, which routes it to the temporal lobe for interpretation.

The function of equilibrium associated with the ear involves the semicircular canals and the vestibule. The *vestibule* is an oval, central portion of the bony labyrinth of the inner ear. It contains two membranous sacs—the *utricle* and *saccule*—which are connected by a small duct. Attached to the vestibule are three semicircular canals composed of a bony and membranous structures.

Each is arranged at right angles to the other two, which provides a three-dimensional configuration when integrated. The semicircular canals are attached by enlargements called ampulla. The vestibular branch of the vestibulocochlear nerve branches into three, connecting to the *ampulla*, the *saccule*, and *utricle*. Static equilibrium relates the head's location with respect to the force of gravity. Any movement of the head translates into linear acceleration, like that experienced in stopping or riding an elevator. Dynamic equilibrium is rotational acceleration. All equilibrium changes involve the detection by the saccule, utricle, and semicircular ducts, which are collectively called the *vestibular apparatus*. The saccule and utricles have two small areas that are thickened. They are called the maculae and are arranged at right angles, detecting linear changes in velocity (acceleration). Maculae consist of hair cells with *stereocilia* and otolithic membrane, which is a thick layer of glycoprotein with calcium carbonate crystals. Inertia results in the bending of the hair cells, creating the impulse we experience as linear velocity changes. In the semicircular duct's ampulla is a small elevated *crista*, which consists of hair cells and supporting cells. These hair cells are covered with a gelatin-like mass (the *cupula*) without the calcium carbonate crystals. The bending of the hair cell bundles creates the impulses associated with rotational velocity changes.

Acting together, the saccule, utricle, and semicircular ducts provide positional information about the movement of the body and head within the environment.

Sensory conflict between the equilibrium and visual sensations results in motion sickness, which is a sensory disorientation issue.

CHAPTER EIGHT

THE CARDIOVASCULAR SYSTEM

ORGANIZATION AND FUNCTION OF BLOOD, HEART, AND VASCULAR ELEMENTS

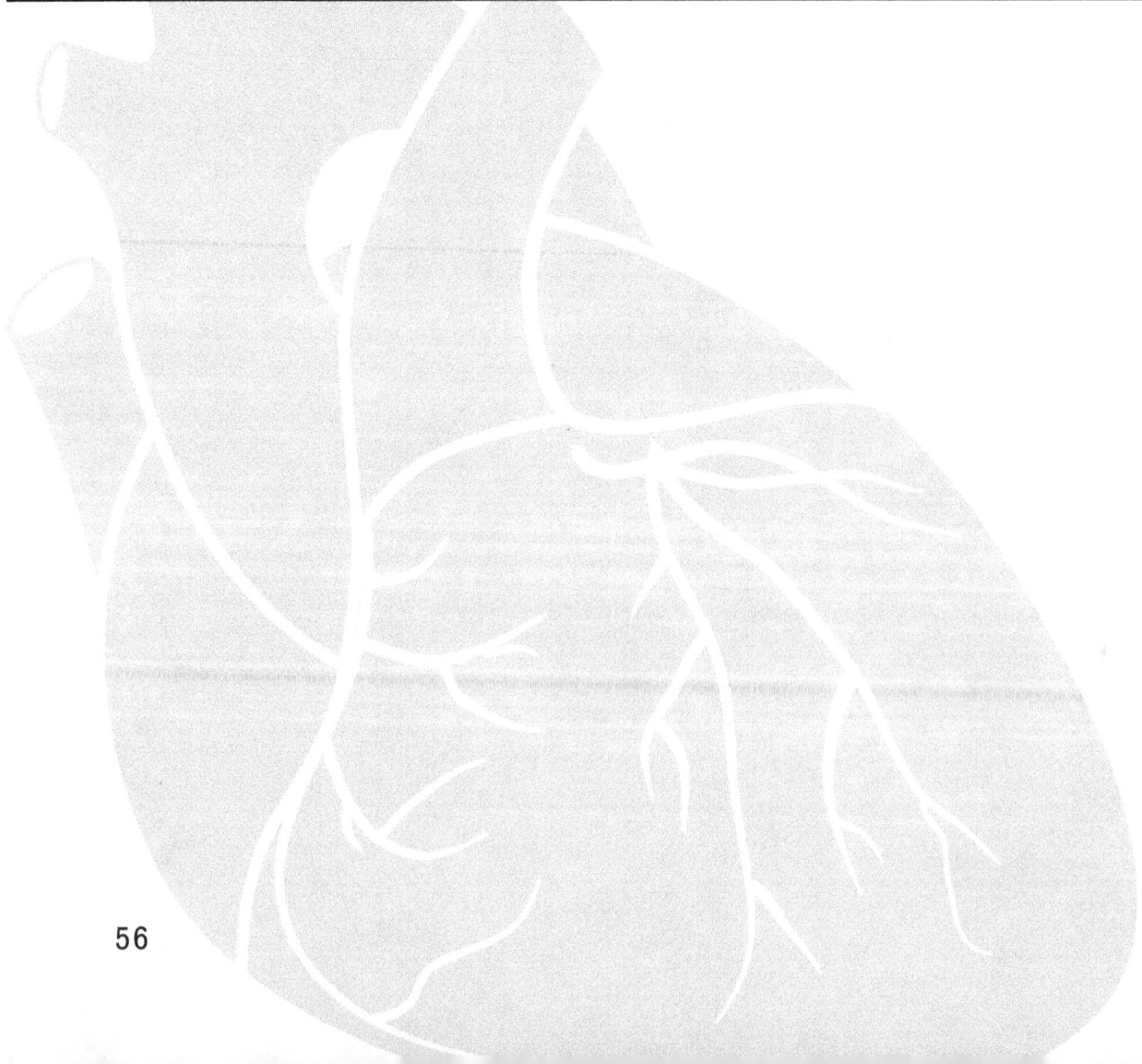

Upon completion of the chapter readers will have a basic knowledge of the cardiovascular system and its physiology:

1. The composition of blood and the functions of its various components
2. The anatomy, organization, blood circuits, physiology, and function of the heart
3. Vascular anatomy and the patterns of circulation in the body

The circulatory system transports fluids, nutrients, and wastes throughout the body. It is composed of two major divisions: the cardiovascular system, consisting of the blood, heart, and blood vessels; and the lymphatic system, consisting of lymph, nodes, spleen, and lymphatic vessels. These systems work in concert to manage the transport of materials, protect the body, and remove wastes.

Cardiovascular System

The cardiovascular system transports oxygen from the lungs to the tissues, and carbon dioxide from the tissues to the lungs; metabolic wastes from the tissues to the kidneys or lungs for elimination; nutrients from the digestive system to various locations in the body; heat from deep tissues and muscles to the integument for elimination; and hormones from the endocrine system to locations of receptors in the body. It also supports stem cells, which form blood components, and transports the cells throughout the body. Protection is another role of the cardiovascular system, with platelets initiating clotting to minimize blood loss; macrophages and microphages ingesting cell debris and foreign cells to prevent infections; stimulating the inflammatory processes to inhibit spread of infective agents; and producing antibodies to neutralized toxins, marking foreign cells for destruction and monitoring the cells of the body to identify and remove atypical cells. The system is involved in regulation by the absorption and release fluids to stabilize fluid distribution; and manages pH with a series of inorganic and organic buffers in the cellular fluid, blood, and cells.

The cardiovascular system functions with the blood as the liquid media, pumped by the heart through vessels, arteries, capillaries, and veins throughout the body to ensure all tissues are supported physiologically.

Blood

Blood is the liquid component of the cardiovascular system. It is composed of a fluid matrix, *plasma*, which is approximately fifty-five percent of the volume, and formed elements, which make up approximately forty-five percent of the volume. These values vary by age, gender, environment, and health status. The above are the approximate normal values. The formed elements consist of erythrocytes (red blood cells), which are the most numerous (four to six million cells in a microliter of blood); platelets (cell fragments with 130,000 to 360,000 fragments in a microliter of blood); and leucocytes (white blood cells normally ranging from 5,000 to 10,000 cells in a microliter of blood).

Plasma

Plasma is a complex mixture of water, proteins, nutrients, electrolytes, nitrogenous wastes, hormones, and dissolved gases. Some of the proteins are soluble and, when stimulated, clot. If a sample of blood is permitted to clot, the fluid that remains is called serum (sera), which is the matrix that no longer has the soluble protein fibrinogen.

Proteins in plasma are involved in clotting, defense of the body, and the transport of iron, copper, lipids, and hydrophobic hormones. A sample of the plasma contains six to nine grams of protein in each deciliter of blood. These proteins are divided into *albumins*, which are the smallest proteins and act as buffers, transport solutes, and are involved in the maintenance of viscosity and osmolarity, which effects blood volume, flow, and pressure; *globulins*, which are involved in transport, clotting, and immunity; and *fibrinogen*, which when stimulated becomes insoluble *fibrin*, which forms the framework of clots. Other proteins in the plasma act as enzymes. The liver produces about four grams of proteins that enter the plasma each hour. The liver does not produce globulins, these are produced by plasma cells descended from lymphocytes.

Nitrogenous waste consists of urea and free amino acids. Nitrogenous waste is produced by the breakdown of proteins, which release ammonia as a by-product, which is toxic to the body. The liver converts the ammonia to urea, which is transported to the kidneys, with a portion being retained in the blood to aid in balancing the blood volume.

Nutrients in the plasma include glucose, amino acids, fatty acids, phospholipids, vitamins, and cholesterol (manufactured by the liver). Small amounts of oxygen are in simple solution, with a larger amount of carbon dioxide being contained as carbonic acid or bicarbonate acting as a buffer, or in simple solution. Nitrogen is contained in the plasma in a saturated condition it has a role in the lungs as a vasodilator.

Electrolytes contained in the plasma are dominated by sodium, about ninety percent of the cations, with potassium and calcium following. These electrolytes are important in the control of osmolarity of the blood, which contributes to the management of blood volume and pressure.

Formed elements

Stem cells located in the spongy bone's red marrow produce the formed elements by a process *hemopoiesis*. These cells are formed in the embryo's yolk sac as blood islands, becoming the stem cells that migrate into the embryo and inhabit the red bone marrow, liver, spleen, and thymus. Shortly after birth, the liver and spleen cease producing erythrocytes. The spleen and thymus continue to be involved in the production of lymphocytes. There are two primary types of stem cells: the *myeloid stem cells* produce the erythrocytes, megakaryocytes (which give rise to platelets), and leucocytes, except lymphocytes which are produced by *lymphoid stem cells*. The lymphocytes are produced by the lymphoid hemopoietic pathway which differs from those produced by the myeloid hemopoietic pathway.

Erythrocytes

Erythrocytes function in the transport of oxygen to the tissues, and the management and transport of carbon dioxide in the body. The average erythrocyte is seven to eight microns in diameter, and approximately two microns thick at the margins. In the mature form, the erythrocyte in the peripheral circulation lacks a nucleus and mitochondria, which creates a biconcave shape appearing as a donut-like configuration. The plasma membrane is composed of glycoproteins and glycolipids, which reside on the exterior portion of the membrane and define the blood type and Rh factor. The interior of the cell's cytoplasm contains about thirty-three percent hemoglobin, which is capable of transporting four O_2 molecules on each hemoglobin molecule. Hemoglobin is involved in promoting the formation of carbonic acid from carbon oxide and water by the action of *carbonic anhydrase*, to create the primary inorganic buffer in plasma. The hemoglobin also transports about five percent of the carbon dioxide as carbaminohemoglobin.

Hemoglobin is composed of four protein chains or globulins, two of which are alpha chains, and two are beta chains, attached to a central heme group containing a ferric iron core to which will attach the four diatomic oxygen molecules in the normal adult. Fetal hemoglobin differs in that the beta chains are replaced by gamma chains, which are more aggressive in binding and transporting oxygen. Shortly after birth the hemoglobin is converted to the adult form, with the beta chains replacing the gamma chains, and the number of erythrocytes are reduced. The ferric iron in the hemoglobin is largely recycled as the erythrocytes are decomposed by macrophages. Some iron is lost in the process, about 1.7 milligrams a day in women and 0.9 milligrams a day in men. This loss of iron must be made up from dietary sources. The pH of the stomach converts the iron so that it is transported as *gastroferritin* to the small intestines, where it is absorbed and binds with *transferrin*. The transferrin transports the iron to the liver, where some is stored; or to bone marrow, where it is used to form hemoglobin; and to other tissues. Muscle tissue uses the iron to manufacture myoglobin to store oxygen, and other cells use the iron in the electron transport chain in the production of energy. The excess iron in the liver is stored as *apoferritin*, creating a reserve of iron when the dietary sources are insufficient. Hemoglobin in the erythrocytes is broken into the globin fractions, which are further broken into amino acids and recycled by the body. The heme portion of the hemoglobin is first relieved of the iron core which is complexed with *transferrin* for recycling, the remaining components of the heme fraction are converted into *biliverdin* initially, then further converted to *bilirubin* and combined with albumin for transport to the liver. The liver stores the bilirubin in the gallbladder, where it is discharged with the bile into the small intestine. Bacteria in the large intestine convert the bilirubin into urochrome, urobilinogen, and stercobilin.

Erythrocyte life cycle

The average life expectancy for an erythrocyte is 120 days. They are formed in the red bone marrow at a rate of about 2.5 million erythrocytes per second or about 20 milliliters of packed cells each day. It requires between three and five days from the initial division of the stem cells until the erythrocyte is mature and departs the bone marrow. The stem cells become erythrocyte colony–forming units, which are then stimulated by the hormone *erythropoietin* to become erythroblasts, which multiply and synthesize hemoglobin. Once this has completed, the nucleus shrivels and along with the mitochondria is discharged from the cells, which have become *reticulocytes*. Normal peripheral blood samples may contain 0.5 percent to 1.5 percent of the erythrocytes as reticulocytes. In a blood loss, the number of reticulocytes increase, as the cells are leaving the bone marrow more rapidly. Hypoxemia results in an increase of erythropoietin production, stimulating increased erythrocyte production with a lag of three to four days. As the erythrocytes move through the vascular system, the cells are forced through capillaries, stressing the cytoskeleton and the cell membrane, causing damage that cannot be repaired without mitochondria and a nucleus. As the erythrocytes proceed to the spleen, many are phagocytized for recycling of the iron and globulins.

Disorders of the erythrocytes include *anemia*, which reduces the capacity of oxygen transport; *polycythemia*, which increases the viscosity of the blood; sickle cell disease; and *thalassemia*, which results in deformed erythrocytes. Anemia may result from any of these causes. If the production of erythropoietin is inadequate or the bone marrow is unable to produce adequate hemoglobin, the number of erythrocytes and the hemoglobin are reduced, which results in less oxygen transported to the tissues. Excessive blood loss (*hemorrhagic anemia*) because of injury, disease, or surgery results in the reduction of erythrocytes and inadequate oxygen transport. The rapid destruction of erythrocytes (*hemolytic anemia*) by systemic bacterial infections or toxic conditions reduces the erythrocytes and negatively impacts oxygen transport. If kidney failure is occurring, the production of erythropoietin results. Insufficient iron in the diet (*iron deficiency anemia*), and insufficient vitamin B12 (*pernicious anemia*) impacts the ability to absorb the iron in the small intestine. A decline in erythrocyte production (*hypoplastic anemia*) or failure to produce erythrocytes (*aplastic anemia*) contribute to the anemic condition, where the lack of oxygen transport can lead to heart and kidney necrosis, as a reduction in the osmolarity of the blood with associated edema creates a higher workload on the heart, which can result in failure.

Polycythemia is frequently associated with cancer of the bone marrow, which is primary polycythemia and results in hematocrits of eighty percent, and erythrocyte counts of eleven million per microliter of blood. Dehydration creates secondary polycythemia with erythrocyte counts between six and eight million per microliter, and can be more easily corrected by fluid intake. Polycythemia is encountered in individuals with emphysema, those who live at altitudes above 8,000 feet, and those who undertake extreme training. The increased erythrocytes associated with polycythemia increase blood pressure, blood volume, and viscosity, imposing additional workload on the heart and vascular components, thus increasing the risk of emboli and stroke.

Sickle cell disease and thalassemia are hereditary defects of the hemoglobin in which *glutamic acid* is replaced by *valine* in the beta chain, creating a folding of the cell membrane and reducing the transfer of oxygen into the erythrocyte. Sickle cell disease occurs in individuals who are homozygous for the trait. This trait is found in individuals whose ancestors were successful in regions with endemic malaria.

Blood type

An individual's blood type is based on the configuration of glycoproteins located on the surface of erythrocyte. The common typing is based on the Lancefield grouping of ABO, which is based on the antigenic properties of the erythrocyte and the corresponding antibodies in the plasma. An individual whose blood is Type O lacks antigenic properties on the erythrocyte and has antibodies for both A and B cells in the plasma. An individual whose blood is Type A has the antigenic properties for A in at least one of the two antigenic sites on the erythrocyte, and the antibodies for B in the plasma. An individual whose blood is Type B has at least antigens for B on one site and antibodies for A in the plasma. An individual with Type AB has A antigen on one site and B on the other, and the plasma lacks antibodies for A and B. The other major factor in typing is the Rhesus factor, which is listed as D. There are two antigenic sites on the erythrocyte. If either of has D antigen then the individual will be Rh positive. This is a concern for young women who are Rh negative and will develop antibodies against Rh positive if they have a fetus who is Rh positive. This can lead to severe hemolytic conditions in both the mother and subsequent fetuses (*erythroblastosis fetalis*), creating severe anemia. ABO conflicts can result in anemia and heart failure due to the development of emboli in the blood.

Leucocytes (white blood cells)

The leucocytes are divided into two major subgroups based on the characteristics of the cytoplasm and its staining character. The cells whose cytoplasm contains large granules are grouped as *granulocytes*, and those without conspicuous granules as *agranulocytes*. In peripheral blood samples the granulocytes are the most numerous, making up more than half of those cells present.

Granulocytes

Neutrophils are the most numerous granulocytes in peripheral blood. The contained granules are small, staining reddish violet to lilac in color, with a nucleus that is lobed in more mature cells and horseshoe shaped in immature cells. The cell are 1.2 to 1.8 times the size of the erythrocytes (9 to 12 microns in diameter), and make up between sixty to seventy percent of the white cells in a peripheral blood sample. They are microphages and capable of releasing antimicrobial chemicals.

Eosinophils numbers are highly variable, ranging between two to four percent of the white cells in the peripheral blood sample. They have large coarse granules that stain orange or rosy red. The cells are 1.5 to 2 times the size of erythrocytes (10 to 14 microns in diameter) and have a bilobed nucleus. The numbers vary with time of day, season of the year, and other physiological and disease factors. They are microphages that phagocytize anti-antibody complexes, allergens, and inflammatory chemicals. They are capable of releasing enzymes that weaken or destroy intestinal parasites.

Basophils are the least numerous granulocytes, making up 0.5 to 1.0 percent of a peripheral blood sample. Their cytoplasm is filled with granules that stain dark purplish blue or black, obscuring the S-shaped nucleus. The cells are 1 to 1.4 times the size of erythrocytes (9 to 10 microns in diameter). The numbers of basophils increase in various viral diseases, diabetes and polycythemia. The cells secrete histamines, stimulating an increase in blood flow. They are associated with mast cells, which line blood vessels and secrete heparin, promoting increased white cell mobility and preventing clots.

Agranulocytes

Lymphocytes are the most numerous of the agranulocytes, being twenty-five to thirty-three percent of a peripheral blood sample. The cells range in size from the size of an erythrocyte to 2.2 times the size (5 to 17 microns in diameter). The cytoplasm stains blue when present, with a round or ovoid nucleus. The subtypes of lymphocytes are not distinguishable morphologically. The majority of the lymphocytes in the body are not in peripheral circulation but in the lymphatic system in the lymphoid structures.

Monocytes make up three to eight percent of a peripheral blood sample. These are frequently the largest whites cells encountered, being 1.5 to 2 times the size of erythrocytes with an ovoid or horseshoe-shaped nucleus. The cytoplasm has very fine granules that stain light blue, many times with vacuoles. The monocytes move out of the peripheral blood, becoming macrophages in the surrounding tissues. They phagocytize pathogens, atypical cells, and cellular debris, and present antigens to activate the immune system.

Platelets

Platelets are cell fragments produced from *megakaryocytes*. They stimulate clotting to stop breaks in blood vessels. A peripheral blood sample contains 130,000 to 400,000 platelets per microliter, with the average being 250,000. Each platelet is between two to four microns in diameter. Each contains lysosomes, microtubules, mitochondria, and microfilaments. They are capable of amoeboid movement. Platelets secrete vasoconstrictors to reduces the lumen of blood vessels, reducing blood flow; adhere to one another, forming a plug in smaller vessels; secrete factors that stimulate conversion of *fibrinogen* into *fibrin*; secrete factors that attract neutrophils, monocytes, and fibroblast to the site of a break; stimulate repair of breaks; and secrete enzymes that eventually dissolve the clot. Megakaryocytes reside in the bone marrow and are up to 150 microns in size. They are stimulated to produce platelets by *thrombopoietin* by a process of thrombopoiesis. The spleen stores twenty-five to forty percent of the platelets, which survive about ten days and are then phagocytized.

Clot formation

The break of any blood vessel stimulates mast cells in the lining of the vessel to release factors that attract platelets and basophils to the break site. The platelets release serotonin, which stimulates a *vascular spasm*, reducing blood flow into the area. The platelets adhere to one another when exposed to the collagen and rough surfaces of the break, drawing the surfaces together, plugging the break in smaller vessels and attracting additional platelets to the area. The damaged tissue of the vessels and surrounding tissue release thromboplastin, which combines with calcium to creates a coagulation cascade, stimulating the conversion of fibrinogen into fibrin, which attaches to the platelets in the break and form a net-like trap for erythrocytes. The fibrin contracts after a period of time, creating a tighter plug at the break site. This requires about thirty minutes, and is called clot contraction. Platelets and endothelial cells release growth factors that stimulate fibroblasts and smooth muscles to begin the process of repair. Once the repair is completed, fibrinolysis occurs, dissolving the clot.

Heart

The heart is the pump that creates the movement of blood throughout the body. It is located in the middle of the thorax (called the mediastinal space) with a tilt to the left and the broad superior portion with major vessels impinging on the left lung. The apex of the heart rests on the diaphragm. It weighs about 300 grams and is the size of an individual's fist. It is enclosed in the *pericardial sac*, divided into a visceral and parietal portion containing between five and thirty milliliters of pericardial fluid that lubricates the space. Reduction of the pericardial fluid would create friction with each heartbeat; increasing the fluid would reduce the refilling of the heart with blood, reducing the cardiac output—a condition called *cardiac tamponade*.

The heart wall is composed of three layers. The outermost layer is a thin *epicardium*, a visceral serous membrane with included adipose tissues over the primary cardiac vessels. The middle layer is the *myocardium*, composed of cardiac muscle arranged in a spiral around the heart and the fibrous framework of the heart. The inner layer is the *endocardium*, which is composed of the endothelial lining and associated thin layer of areolar tissue covering the valves and being continuous with the endothelium of the blood vessels. The fibrous framework of the heart consists of collagenous and elastic fibers concentrated into fibrous rings between the chambers and around the valves. This framework functions to provide support for the valves and major vessels; provide anchoring for the muscle cells; insulate the atria and ventricles, electrically controlling the stimulation of the myocytes; and aid in the refill of the heart during diastole.

The heart is divided into four chambers, with a left and right of each. The receiving chambers are the *atria*, and the muscular ejecting chambers are the *ventricles*. The atria are located superior and posterior on the broad base. They are thin walled and receive blood from the systemic and pulmonary circulation. The atria delivers blood to the ventricles through *atrioventricular valves*, which close when the ventricles contract. The ventricles are located inferior and anterior near the diaphragm. There are a series of grooves (*sulci*) that mark the boundaries of the chambers and contain vessels that distribute blood to the heart itself. Internally, the atria are separated by a septum into a right and left. The right atrium has a series of ridges, the *pectinate muscles,* which aid in the movement of blood into the right ventricle through the *tricuspid valve*. In both ventricles a series of ridges called *trabeculae carneae* aid in creating a turbulent condition to mix the blood during the ventricular contraction.

All of the valves in the heart are one-way (check) valves, permitting blood flow in only one direction unless the valve fails. The atrioventricular valves, *tricuspid* (between the right atrium and ventricle), and *bicuspid* (*mitral*) (between the left atrium and ventricles) are normally open during *diastole*, permitting blood to flow into the ventricles. The atrioventricular valves are composed of a series of leaflets (cusps), which are attached to the *chordae tendineae* and *papillary muscles* in the ventricles. These leaflets are endothelial tissue. The *semilunar valves*, the *pulmonary semilunar valve* from the right ventricle and the *aortic semilunar valve* from the left ventricle, isolate the ventricles from the pulmonary and systemic circulation, preventing backflow into the ventricles. The semilunar valves are shaped like cups, which are directed away from the ventricles and normally closed by the back flow of the blood once the ventricles' ejection of blood decreases. In order to open the semilunar valves, the ventricles must increase the pressure greater than the back pressure on the valves. The right ventricle must create pressure between 15 and 25 mmHg to open the pulmonary semilunar valve and eject blood through the lungs for oxygenation. The left ventricle must create pressure between 85 and 100 mmHg to open the aortic semilunar valve and begin ejection of blood into the systemic circulation. Common problems with valves can be expected to occur on the left-side valves where the bicuspid valve leaks, creating a condition called *prolapse*; or the valve may become stiff, which is called *stenosis*. The aortic semilunar valve may stiffen or develop extra tissue (scarring), preventing complete closure, which can result in major cardiovascular problems.

Blood flows through the heart in a figure-eight pattern. The blood is received from the systemic circulation by the *superior* and *inferior vena cava* into the right atrium by changes in the ventricle, thoracic pressures, and muscular pumping. The majority of the returned blood flows directly into the ventricle. The atrium is stimulated and contracts, topping off the blood in the ventricle and slightly stretching the ventricular muscle. The right ventricle begins contracting, the blood forces the tricuspid valve closed, and contraction continues until the pulmonary semilunar valve opens when the contraction accelerates, ejecting blood into the pulmonary circuit. The left atrium receives blood from the pulmonary veins with the majority (about eighty percent) going directly to the left ventricle. The left ventricle begins to contract at the same time as the right ventricle. The bicuspid valve is forced closed. The ventricle continues contacting until sufficient pressure is generated to overcome the pressure acting on the aortic semilunar valve, which gradually opens. Once the valve opens, the contraction rate of the left ventricle accelerates, ejecting a larger volume of blood and creating a *wedge pressure* in the systemic circuit, which creates the *systolic peak* of the blood pressure. Once about fifty to sixty percent of the blood content of the ventricle has been ejected, the pressure decreases until the aortic semilunar valve is forced closed. The opening and closing of the valves create the heart sounds.

The third circuit is the *coronary circulation*, which is the blood supporting the heart's physiology. The heart receives approximately 250 milliliters of blood per minute, which is about five percent of that pumped from the coronary arteries, which branch off the ascending aorta just above the aortic semilunar valve. The *left coronary artery* travels through the coronary sulcus and divides into two branches: the *circumflex* and the *anterior interventricular arteries*. The anterior interventricular branch continues through the interventricular sulcus to the apex of the heart, extending further to the posterior of the heart where it joins with the posterior interventricular branch. These vessels supply two-thirds of the blood to the ventricles and the interventricular septum. The circumflex branch follows the coronary sulcus to the left side and gives rise to the *left marginal branch*, furnishing blood to the left ventricle and atrium. The *right coronary artery* supplies blood to the right atrium, continuing through the coronary sulcus and branching into the *right marginal branch*, which supplies the apex of the heart, lateral atrium, and ventricle; and the *posterior interventricular branch*, supplying blood to the posterior portion of both ventricles and joining the circumflex and interventricular branches of the left coronary artery. About twenty percent of the blood in the coronary circulation returns directly into the right atrium through multiple small veins, while the majority is collected and returned through the *coronary sinus*. The blood entering the coronary sinus is collected from the *posterior interventricular (middle cardiac) vein*, draining the posterior apex of the ventricles and the *left marginal vein* collecting from the left margin of the heart.

Cardiac muscle physiology

Heart muscle is rhythmic, averaging seventy-five beats per minute in a resting adult. The heart is myogenic, generating the stimulus for contraction within the heart itself at the *sinoatrial node*. Heart stimulation and rhythm can exist with no external nerve supply, but is unable to respond to stresses. Both sympathetic and parasympathetic nerves innervate the heart, modifying the strength and rate of the contractions. The sympathetic pathway can stimulate the rate to increase up to 230 beats per minute under stress, while parasympathetic pathways can slow the heart to 20 beats per minute or even short-term stoppages. The sympathetic pathway originates in the lower cervical and upper thoracic spine, travelling through the cervical ganglia then through the cardiac nerves to the ventricular myocardium. The parasympathetic pathway innervates the sino-atrial node and atrioventricular nodes through the vagus nerve, slowing the heart rate. Without the effect of the vagus nerve, the resting heart rate would be about a hundred beats per minute, while the vagal tone reduces it to between seventy and eighty beats per minute.

Cardiac muscle cells (myocytes) are autorhythmic, depolarizing in a regular pattern. The *sinoatrial node* is located in the right atrium near the superior vena cava, initiating the stimulus of the contraction processes. The sinoatrial node depolarizes, stimulating the atria muscles, passing the stimulation through the right and left atrial muscles, and stimulating the atrioventricular node. The *atrioventricular node* is located near the base of the tricuspid valve, adjacent to the atrioventricular septum. The atrioventricular node delays the impulse, permitting the final filling of the ventricle and then discharging, stimulating the conduction of the stimulating impulse through the *bundles of HIS* and the *right* and *left bundles branch*es to stimulate the ventricular muscle cells by the *Purkinje fibers*. The bundles pass the impulse through the interventricular septum to the apex and spread the stimulation up the external myocardium.

Unlike the skeletal muscle cells, the cardiac muscle cells do not have to be directly stimulated to contract. They are capable of passing the stimulation cell to cell by the tight gap junctions created by the *intercalated disc*. The cardiac myocytes are striated with a centrally located nucleus. The cells are between fifty and one hundred microns long and ten to twenty microns thick. The cells are branched, connecting with several other myocytes capable of passing the electrical stimulus from cell to cell rapidly. The cardiac myocytes have numerous large mitochondria and store oxygen in myoglobin. The cells obtain a large amount of calcium ions from extracellular sources to aid in the contracting process. Cell repair and regeneration is very limited. The cells are rich in glycogen. The cardiac muscle cells function aerobically, using fatty acids for sixty percent of their energy and only using glucose for thirty-five percent of their energy requirements. Cardiac muscle deprived of oxygen for a period in excess of ten minutes will begin to die.

The normal rhythm of the heart triggered by the sinoatrial node is called *sinus rhythm*. It ranges from seventy to eighty beats per minute, but the beats may range from sixty to one hundred, which is not uncommon. The sinoatrial node is impacted by hypoxia (low oxygen), imbalances in electrolytes, caffeine, nicotine, and various medications, all of which can create premature ventricular contractions or extra beats, some caused by ectopic foci attempting to supplant the sinoatrial node. Under some conditions the major ectopic foci is the atrioventricular node, which would slow the rhythm to forty to fifty beats per minute and is called *nodal rhythm*. Other ectopic foci can result in rates of twenty to forty beats per minute, which impacts long-term survival. *Arrhythmias* (atypical rhythms) frequently are due to dysfunctions in the conduction of the impulses, such as a blockage of one of the bundles. Damage to the atrioventricular node would result in a total heart block and the atria and ventricles would beat without coordinated function, which would not support life long term.

Physiologically, the sinoatrial node fires at regular intervals. Its action potential is unstable in the resting state. Upon repolarization the action potential reaches –60 mV and it begins to drift upward, slowly depolarizing by a slow inflow of sodium ions and loss of potassium ions until it reaches a threshold of –40 mV, which causes fast calcium channels to open, pushing the potential to 0 mV, where the potassium channels open, initiating the repolarization. This cycle occurs approximately every 0.8 of a second, establishing the heart rate of about seventy-five beats per minute. The impulse generated by the sinoatrial node excites the muscle cells of the atria, taking approximately fifty milliseconds to reach the atrioventricular node. The atrioventricular node slows the impulse for about a hundred milliseconds. The impulse is conducted to the ventricle myocardium through the bundles and Purkinje fibers at a rate of four meters per second to stimulate the entire ventricular myocardium in approximately 200 milliseconds after the initial sinoatrial node firing. The papillary muscles are stimulated earlier, placing tension on the chordae tendineae prior to the ventricular contraction, which proceeds in a spiral fashion toward the semilunar valves. The cardiac myocytes remain depolarized for 200 to 250 milliseconds and have an absolute refractory period, preventing the tetany and summation found in skeletal muscle.

The action of the stimulation of depolarization and repolarization of the atrial and ventricular muscles is detected and measured by the electrocardiogram. This measures the electrical changes on the surface of the skin stimulated by the electrical activity of the heart. Electrodes are placed on the surface of the skin to detect electrical changes. The P wave is the electrical change at the surface stimulated by the depolarization of the atrial muscles prior to atrial systole; it is between ninety and one hundred milliseconds in duration. The next change is the QRS complex,

which indicates the stimulation and depolarization of the massive ventricular muscle mass. Atrial repolarization occurs during this event and is masked by the large amount of muscle involved. The T wave is the repolarization of the ventricle muscle mass. This entire event lasts 540 to 700 milliseconds in duration. The atria contracts from the area around the superior vena cava toward the ventricles, beginning about fifty milliseconds after the beginning of the P wave and lasting through the interval between the P wave and the Q deflection. The ventricles begin to contract shortly after the R wave peak, beginning at the apex and contracting against closed valves, called isovolumetric contraction, where no blood is ejected. The ejection of blood from the ventricles begins at about the S wave and continues until about the beginning of the T wave, at which time the semilunar valves close and the ventricles continue to contract into the T wave without ejecting blood.

The cardiac cycle consists of a complete contraction and relaxation of the chambers. During the contractions and relaxation the pressures change, affecting the blood flow, which creates sound as the valves open and close. Auscultation (listening for heart sounds) provides two or three distinct sounds, which are identified as S1, S2, or S3. The loudest and longest is the S1, referred as *lubb*, which is followed by the softer and sharper S2 or dub sound. These sounds are created when the blood flow is turbulent as the valves open and close. S3 sound is rare in adults over thirty years of age. The cardiac cycle begins with the filling of the ventricle during diastole, when the ventricles expand after contraction with pressure dropping below the pressure in the atria. The atrioventricular valves are open, so this initial filling is rapid. After about one-third of the ventricle is filled, the next third is slower, due to gravity and thoracic pressure changes. The final filling of the ventricle is due to the atrial contraction, which stretches the ventricle slightly and contributes about 40 milliliters to the *end diastolic volume* of about 130 milliliters. Once the ventricle is filled and stimulated, it begins to contract at the apex, closing the atrioventricular valves by blood surging against the cusps (leaflets). This period generates the S1 sound mainly from the left ventricle and is the *isovolumetric contraction* without ejection of blood. Ventricular ejection begins when the internal pressures of the ventricles reach or exceed the back pressures on the semilunar valves, usually 10 to 25 mmHg for the pulmonary circuit and 80 to 100 mmHg for the systemic circuit. The ventricles eject about seventy milliliters of blood, which is the *stroke volume*. The pressure and flow decreases with the semilunar valves closing, generating the S2 sound. The blood remaining in the ventricle provides a reserve for more vigorous activity and is called the *end systolic volume. Isovolumetric relaxation* occurs as the ventricular diastole begins, with the expansion of the ventricle by the fibrous framework. Atrial systole lasts approximately 100 milliseconds, and ventricular systole last about 300 milliseconds. All chambers operate in concert; diastole in all chambers lasts about 400 milliseconds, while the entire cardiac cycle occurs over a period of about 800 milliseconds, or a resting heart rate of seventy-five beats per minute.

The importance of the cardiac cycle is the *cardiac output* (CO), or how much blood the heart moves around the body per minute. The heart rate and the stroke volume multiplied together is the cardiac output (HR X SV = CO). The case used of 70 milliliters stroke volume and heart rate of 75 beats per minute yields a cardiac output of 5.25 liters per minute. This would mean that a person with a normal blood volume between four and six liters would move all or the majority of the blood around the body each minute. Highly conditioned individuals are capable of cardiac outputs of thirty-five liters per minute, while normal individuals are capable of cardiac outputs of twenty-one liters per minute during vigorous activity. Persistent heart rates over one hundred beats per minute because of stress, anxiety, fever, or heart disease create a condition called *tachycardia*, which can also be encountered in heat stress and significant blood loss. Persistent heart rates below sixty beats per minute can occur during sleep or in well-conditioned endurance athletes, and is called *bradycardia*, which may also be encountered in an individual who is hypothermic.

Autonomic controls of the heart occur through the cardiac centers located in the medulla oblongata. There are two neural pools. One of which, the *cardiac acceleratory center*, controls the heart rate, stimulating the sympathetic pathway to the sinoatrial node to increase the heart rate, increasing the cardiac output and peaking between 160 and 180 beats per minute. This increase occurs by shortening the ventricular diastole from about 620 milliseconds to 140 milliseconds. The other neural center, the *cardiac inhibitory center*, signals through the parasympathetic pathway of the vagus nerve, increasing the vagal tone to slow the heart rate. These centers are influenced by proprioceptors, baroreceptors, and chemoreceptors, which generate stimulation, enabling the body to cope with changes it is experiencing.

Vascular System: Blood Vessels

The vascular portion of the cardiovascular system is composed of the blood vessels, which transport and distribute oxygen and nutrients throughout the body and return the carbon dioxide and other wastes to other locations for removal. The arteries transport blood from the heart to various locations, becoming smaller in size the further they are from the heart until only the endothelial lining continues as capillaries, which connect to small venules (veins), which become larger the closer they get to the heart. The vessels consist of three distinct layers, called tunics. The inner layer (*tunica interna* or *intima*) is the endothelial lining of the blood vessels, composed of simple squamous epithelial tissue supported by a basement membrane and thin loose layer of connective tissue; this layer provides a selectively permeable barrier and is able to secrete chemicals to constrict or dilate the vessels. The middle layer (*tunica media*) is the thickest layer of the vessels, consisting of smooth muscle, collagen, and elastic fibers, which vary with the size of the vessel. The outermost layer (*tunica externa*) consists of loose connective tissue, which merges with the surrounding tissues, forming the *vas vassorum* to supply blood to the tunica media.

Arteries

The arteries are strong, resilient vessels that expand with pressure wedges created by the ventricular ejection and contract to maintain flow. The arteries can be divided into three types, based on their size. *Conducting (elastic) arteries* are the largest, represented by the aorta, common carotid, subclavian, common iliac, and pulmonary trunk. These arteries have a thin internal elastic layer between the tunica interna and media; the tunica media is thicker, with alternating thin layers of smooth muscle, collagen, and elastic fibers. These vessels are expanded with the *wedge pressure* of the ventricular ejection and then contract, forcing blood flow to continue forward in the system. The *medium (muscular) arteries* are smaller arteries that distribute blood

to specific organs (femoral, splenic, brachial, and renal). The tunica media is composed of smooth muscle, which composes about seventy-five percent of the wall thickness and has more obvious elastic tissue. The resistant arteries are the smallest arteries; they have little elastic tissue and the tunica media is thicker than the lumen of the vessel. These link to the arterioles with small amounts of smooth muscle, and the tunica externa is reduced. These connect with the capillaries. The major arteries above the heart contain sensory receptors that communicate with the cardiac control centers to regulate strength of contract and heart rate, and stimulate a response in to dilate or contract small aterioles regulating capillary flow. The carotid arteries have a sinus that contains baroreceptors to measure changes of blood pressure, which send signals through the glossopharyngeal nerve fibers to the brain stem; and chemoreceptors, which detect shifts in pH to adjust respiratory activity. In the aorta are aortic bodies, which contain chemoreceptors that detect pH and alter the respiratory activity.

Capillaries

Capillaries are composed of the endothelial lining continuation of the tunica interna and the basement membrane. They are thin walled, narrow at the arteriole attachment, and expand toward the venule attachment. The lumen of the capillaries is about the diameter of the erythrocyte (eight microns). The capillaries permit diffusion of gases (oxygen, carbon dioxide), nutrient release, and waste absorption to and from surrounding tissues. Capillaries are absent in the integument, cartilage, cornea, and lens. There are three types of capillaries. The *continuous capillaries* are the most common, connected to arteriole on one end and venules on the other, and are found in most tissues. The *fenestrated capillaries* have little windows or pores, permitting large molecules to enter and exit the capillaries. These are encountered in the kidneys, endocrine glands, choroid plexuses, and small intestine, where the pores are selective for molecular size. The *sinusoid capillaries* are encountered in the liver, spleen, and bone marrow. They are twisted, with large openings, permitting very large molecules and formed elements to enter the cardiovascular system.

Veins

Veins are capacitance vessels. At any time they contain about fifty percent of the systemic blood. The smallest are venules, which receive blood from capillaries and are composed of the tunica interna with the addition of fibroblasts. The muscular veins are slightly larger, with a thin tunica media and very thicker tunica externa. The next larger veins (radial, ulnar, saphenous) have a complete tunica interna, with folds of the endothelial lining acting as valves to provide directional blood flow with muscle pumping. The larger veins have a thick tunica externa with longitudinal bundles of smooth muscle. There are specialized venous structures, which are thin walled with large lumens acting as sinuses (*coronary sinus* and *dural sinuses* of the brain).

Circulation

Circulatory pathways progress from the heart by arteries to the capillaries, to the veins, returning blood to the heart. There are several locations where the blood goes from capillaries to veins to capillaries. These are called portal systems, found in the renal system, hepatic system, intestinal system, and in the hypothalamus and anterior pituitary. In other areas vessels may merge, which is called anastomosis, providing alternate pathways to serve the organs, as in the circumflex artery of the heart. There are shunts allowing the blood to be rerouted to reduce heat loss.

Blood pressure and flow are rapidly variable due to local controls stimulated by the accumulation of metabolic wastes increasing flow or the removal of wastes reducing flow. Platelets, endothelial cells, and mast cells secrete chemicals that stimulate vessel changes.

Flow in capillary exchange occurs through the endothelial cells, through clefts between the cells, or through pores in the endothelium. The mechanisms that drive these exchanges are diffusion, transcytosis, filtration, and reabsorption. Hydrostatic pressures drive the filtration and reabsorption along with colloid osmotic pressure. If fluid is not recovered from the tissues, a condition called *edema* occurs. Edema results from an increase in capillary filtration rate, reduced reabsorption by the capillaries, or obstruction of the lymphatic drainage. Prolonged edema results in tissue destruction and a condition called *necrosis*.

Venous return is driven by the pressure gradient generated by the central venous pressure, gravity, skeletal muscle pumping, respiratory pumping, and suction created by the relaxation of the ventricles. All of these working together create the venous return to the heart and, if reduced, create cardiac insufficiency.

Shock results from insufficient blood reaching the brain. It can result from inadequate pumping of the ventricles, which results in *cardiogenic shock*; or reduction in the venous return, which is *hypovolemic shock* due to injury, dehydration, obstructed venous return, or venous pooling due to stagnant muscle pumping. *Septic shock* can occur from the production of toxins by pathogenic organisms. *Anaphylactic shock* results from the intense response of the immune system to an allergic stimulus.

LABELING ACTIVITY

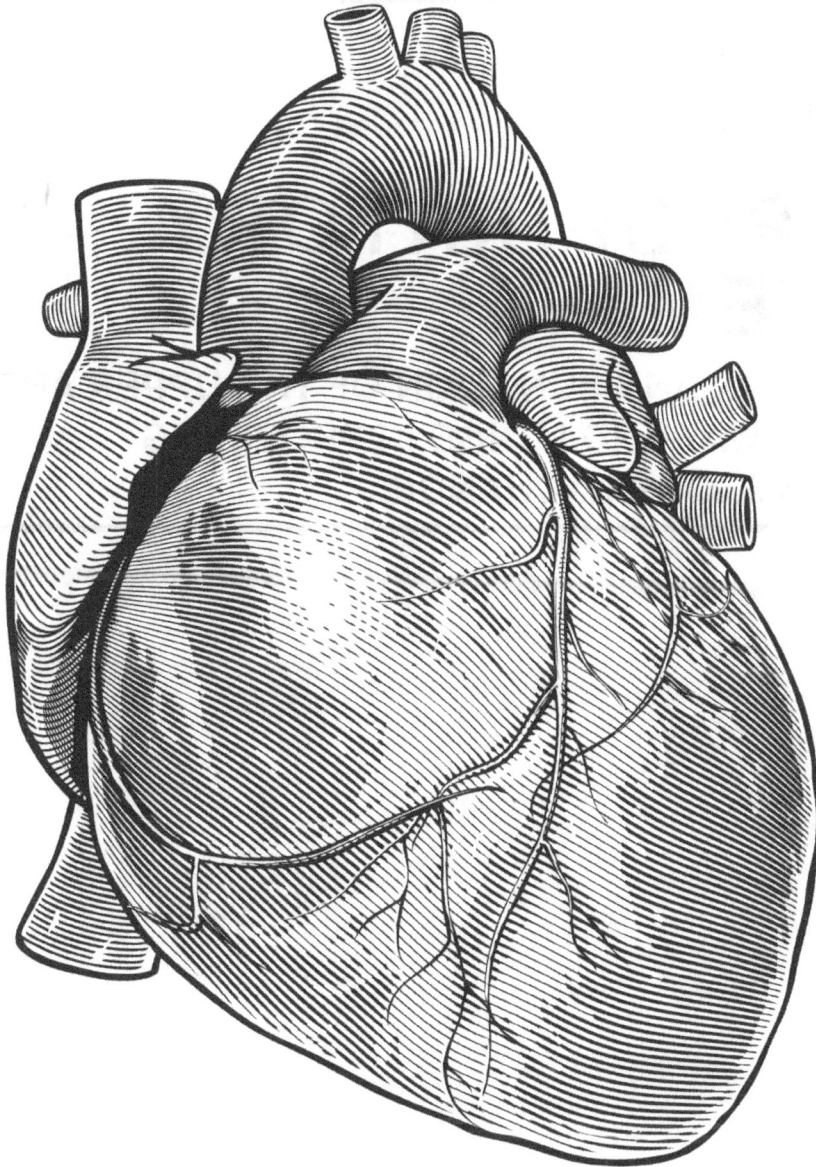

Fig. 8.1 Source: Copyright © 2014 Depositphotos/Lestyan.

CHAPTER NINE

THE RESPIRATORY SYSTEM

ORGANIZATION AND FUNCTION OF THE UPPER AND LOWER AIRWAY AND LUNGS IN PROVIDING GASEOUS EXCHANGE

LEARNING OBJECTIVES

Upon the completion of the chapter readers will have an understanding of the essential knowledge of the respiratory system's anatomy and physiology:

1. Anatomy of the respiratory system and its divisions
2. The physiology of gaseous exchange including the related gas laws
3. Method used to assess pulmonary function and ventilation
4. Impact of low oxygen conditions and physiological response
5. Interactions between the cardiovascular and urinary systems to manage pH

The respiratory system consists of the structures required to conduct air from the exterior into the body (ventilation), the exchange of gases with the blood, and the transport of those gases to the tissues for the metabolic function of the production of energy. The principal gases of concern are oxygen and carbon dioxide. The air is a mixture of gases, primarily nitrogen (78%) and oxygen (21%) with carbon dioxide, and other gases in minor concentrations. Breathing or respiration means the ventilation of the airway, including the lungs; the diffusion of oxygen into the blood in the lungs and carbon dioxide diffusion out of the blood into the lungs; and the transport of oxygen to the tissues, where it diffuses into the cells from the blood, and carbon dioxide diffuses from the cells into the blood. The anatomical structures involved with ventilation are the nose, pharynx, larynx, trachea, bronchi, bronchioles, and the alveoli, which make up the lungs. Besides the gaseous exchange, the respiratory anatomy promotes speech or vocalization, supports the sense of smell, controls the pH by management of the carbon dioxide levels, synthesis of angiotensin II to regulate blood pressure, promotes venous return, and aids in expelling of wastes. The upper respiratory system consists of all structures above the larynx and lower respiratory is the larynx and below.

Anatomy

The *nose* is the entry point of air from the environment. It acts to filter the air, removing large, slow-moving particles at the opening, warming the air and increasing its moisture content. The nasal passages are lined with mucus membrane, which yields moisture and warmth to the air. The moist surface traps particles that pass the nares, which is divided by a septum formed by the vomer into a left and right. The septum is formed by the vomer bone and the septal cartilage covered with mucus membrane. The nasal cavity is formed by the sphenoid and ethmoid bones making up the dorsal roof, the maxilla making up the lateral sides and anterior floor, and the palatine bone making up the posterior floor. Vibrissae (guard hairs) at the entry of the nares block the entry of large particles and insects into the nasal passages. The upper mucosa, containing the olfactory receptors, detect odor in the air. The majority of the nasal cavity is lined with pseudostratified ciliated columnar cells, which are associated with goblet cells, producing mucus that forms the respiratory mucosa. Beneath the mucosa and associated with it is the lamina propria, which is populated with lymphocytes of the immune system and provides triggers for the immune system upon the entry of pathogens in the inhaled air. Large blood supply of the mucosa aids in the warming of the air inhaled.

The *pharynx* connects to the nasal passages and acts as a muscular funnel extending from the posterior nares to the larynx. The pharynx is divided into the *nasopharynx*, connecting to the posterior nares and extending along the dorsal aspect of the hard and soft palate. The *Eustachian tubes* are located in the nasopharynx. Entering particles larger than ten microns are trapped in the lining of the nasopharynx. The *oropharynx* encompasses the space between the soft palate and the base of the tongue, and contains the palatine and lingual tonsils (parts of the lymphatic system). The *laryngopharynx* begins with the joining of the naso and oro pharynx at the level of the *hyoid bone* and connects to the cricoid cartilage of the larynx.

The *larynx* (voice box) is a cartilaginous structure that separates food and liquids from the airway and produces sound by vibration of the vocal folds. The opening of the larynx is guarded by a flap of tissue *(epiglottis)*, which closes over the opening of the larynx when swallowing, routing food and liquids into the esophagus along the vestibular folds. The larynx is composed of nine cartilage segments. The most superior is the *epiglottis cartilage*, supporting the epiglottis. The largest cartilage is the *thyroid cartilage*, which forms the Adam's apple. The *cricoid cartilage* connects the larynx to the trachea on the inferior end. The remaining cartilage segments are small and paired, supporting the soft tissues and vocal folds. These are the *arytenoid cartilage*, which is posterior to the thyroid cartilage; the *corniculate cartilage*, which has horns for the attachment at the upper ends; and the *cuneiform cartilage*, which supports the soft tissues between the arytenoid cartilage and the epiglottis. The epiglottis, corniculate, and cuneiform cartilages are elastic cartilage; the others are hyaline cartilage. These cartilage segments are connected by a series of fibrous ligaments: the thyrohyoid ligament is the broad ligament connecting the hyoid to the thyroid cartilage; the cricotracheal ligament connects the cricoid cartilage to the trachea's intrinsic ligament; and the vestibular and vocal ligament connect the thyroid and arytenoid cartilages. The vestibular fold closes the glottis when swallowing; the vocal folds (cords) are controlled by intrinsic muscles attached to the corniculate cartilage and arytenoid cartilage, which upon contraction twist the folds, while air forced over the folds vibrates, producing sound which then resonates in the pharynx, oral cavity, tongue, nasal cavities, and with lip activity produces intelligible sound (speech).

The *trachea* (windpipe) is a semirigid tube supported by C-shaped rings of hyaline cartilage located anterior to the esophagus. Along the posterior side of the trachea is the trachealis muscle. The lining of the trachea consists of ciliated pseudostratified columnar epithelial cells; associated goblet cells, which produce mucus; and associated basal stem cells. The mucus produced by the goblet cells traps foreign particles and the cilia of the epithelial cells move the material upward toward the larynx and glottis, where the mucus and its content are swallowed, usually at night. This cleanses the airway of debris. The outer layer of the trachea is the adventitia. It blends with other connective tissues in the mediastinum. The bronchi branch from the trachea into primary bronchi at a tracheal cartilage called the carina.

The *bronchi* are supported by hyaline cartilage rings and branch into smaller secondary and tertiary bronchi. The bronchioles continue from the tertiary bronchi as muscular tubes of smooth muscle and cuboid epithelial cells into smaller bronchioles until they become terminal bronchioles, which are attached to the alveoli.

The *lungs* are the composite structure of the bronchi, bronchioles, and alveoli, along with the supporting connective tissues. The lungs are surrounded by the pleural membrane and the costal curve of the lung rests against the rib cage, which is lined by the parietal pleura. Along the medial surface of the lung are slits called *hilum* where the primary bronchi, blood vessels, and nerves enter the lung. The right lung is shorter and thicker than the left and is divided into three segments by fissures. The left lung is taller and narrower, displaced slightly to the left due to the rotation of the heart; it is divided into two segments by fissures. The bronchi branch in the lungs, forming the bronchial tree, which is accompanied by the pulmonary arteries.

The *alveoli* are the functional locations in the lungs. This is where gaseous exchange occurs. The alveoli are composed of thin squamous epithelial cells (alveolar cells) that cover ninety-five percent of the area. The other five percent is covered by cuboidal epithelial cells, which produce a pulmonary surfactant that is a mixture of phospholipids and proteins, preventing alveolar collapse on exhalation and acting to repair the pulmonary epithelium. Numerous macrophages (dust cells) wander through the alveoli, removing debris and phagocytizing loose cells and bacteria, cleaning house.

The *pleurae* is the serous membrane that covers the lungs and lines the cavity around the lungs. The pleural space is lubricated by approximately fifty milliliters of pleural fluid, which reduces friction during the inflation and deflation of the lungs during ventilation and promotes expansion of the lungs, reducing pressure in the alveoli with contraction of the diaphragm. The serous membrane assists in blocking the spread in case of infection. Fluid in excess of approximately one hundred milliliters reduces the capacity of the lung, creating potential problems.

The *diaphragm* is the skeletal muscle separating the thoracic and abdominal cavities. In a relaxed state it is dome shaped and when stimulated it flattens, increasing the thoracic volume and reducing the pressure. This is the primary muscle of respiration, and frequently forgotten when covering the respiratory system.

Pulmonary Physiology

The physiology of the pulmonary system is driven by a series of gas laws. *Boyle's law* is a gas law which identifies that the pressure of any gas will vary inversely (in the opposite direction) proportional to the volume, if the temperature does not change. Boyle's law says that if the volume changes and becomes larger, then the pressure will decrease. This is the gas law that drives the ventilation of the lungs during inhalation and exhalation. *Charles's law* is a gas law that states that the volume of a gas will be directly proportional to the absolute temperature, with the pressure remaining constant. Charles's law says that when inhalation occurs, the volume of a gas will increase as it warms to body temperature, which happens to a small extent as the air moves through the airway. *Dalton's law* is the gas law that states that in a mixture of gases, the total pressure is the sum of the partial pressure of all the gases in the mixture. This is the gas law that governs the gaseous exchange from the alveoli to the blood and tissues. The fourth gas law is *Henry's law*. Henry's law states that the amount of a gas in any liquid is determined by how soluble the gas is in the liquid, and the partial pressure of the gas over the liquid. This is the gas law that explains that carbon dioxide, which is highly soluble in water, is mostly found in the plasma, rather than bound to the hemoglobin in the erythrocytes.

Pulmonary ventilation (breathing) consists of the inspiration (inhalation) and the expiration (exhalation) of the air from the lungs, which forms the respiratory cycle. The respiratory cycle is driven by Boyle's law, with pressure gradients developed by changes in volume of the thoracic cavity by the action of the diaphragm and accessory muscles. As the diaphragm contracts at rest, it increases the volume by moving 1.5 centimeters; when very active, causing deep breathing, it is capable of moving 7 centimeters. In each case, the enlargement of the thoracic cavity decreases the internal pressure below the external pressure, resulting in the mass movement of air into the airway and lungs; this is *inhalation*. As the diaphragm and accessory muscles relax, the volume of the thoracic cavity is reduced, increasing the internal thoracic pressure, resulting in *exhalation*. Under some conditions the thoracic cavity pressures are increased by straining the abdominal muscles and contraction of the diaphragm without mass air movement, called a Valsalva maneuver. This is encountered in childbirth, forced urination, defecation, and when experiencing extreme acceleration.

Control of breathing is normally below the conscious level. The *pneumotaxic center*, located in the pons, controls the rate of breathing, while the centers in the medulla oblongata control the pattern of breathing in the *apneustic center*. The conscious control of the diaphragm occurs through the *phrenic nerve*, which arises in the cervical plexus. An individual can hold their breath, but upon loss of consciousness breathing will resume. Chemoreceptors located in the carotid and aortic sinuses measure the pH (indirectly measuring the carbon dioxide level), feeding stimulus to the respiratory control centers through the vagus nerve to control breathing.

When the diaphragm and the intercostal muscles contract, the thoracic volume increases, reducing the intrapleural (thoracic) pressure from 1.4 mmHg to –6 mmHg, expanding the lungs, and reducing the alveoli pressure to –3 mmHg, creating the gradient for inhalation. At rest approximately 500 milliliters of ambient air moves into the airway; the air expands slightly due to the influence of Charles's law. The non-functional (no gaseous exchange) portion of the airway is approximately 150 milliliters (anatomical dead space) in volume. Only about 350 milliliters of air reaches the alveoli to support resting gaseous exchange. The relaxation of the diaphragm and intercostal muscles causes the thoracic pressure to increase, raising the pressure to 3 mmHg at rest or 30 mmHg with forced breathing. Routinely, our ventilation cycles at about twelve times per minute (10 to 14 at rest) and moves 500 milliliters of air with each cycle, so that approximately six liters are cycled each minute, or 4,200 milliliters ventilate the alveoli. Under more strenuous conditions the rate can increase to twenty-four breaths per minute and the depth to approximately 2.2 to 2.8 liters per breath, supporting the oxygen demand of the body. Deeper breathing and more rapid breathing can rarely be supported for prolonged periods of time due to the fatigue of the active muscles.

Gaseous exchange at the alveoli is driven by Dalton's law. Oxygen is the physiologically active gas involved in diffusion. In the ambient air the oxygen has a partial pressure of 160 mmHg. This partial pressure is reduced by the water, which saturates the air, reducing the partial pressure in the trachea to approximately 148 mmHg. As it reaches the alveoli, the partial pressure of oxygen is reduced by the increase in partial pressure of carbon dioxide, which displaces oxygen to 107 mmHg, creating the pressure driving oxygen across the membrane into the hemoglobin of the erythrocytes. The result of this diffusion is about ninety-eight percent saturation of hemoglobin, with a partial pressure of 80 to 100 mmHg in the arterial blood compared to the eighty to eighty-five percent saturation and a partial pressure of forty to fifty in the venous blood. Changes in altitude effect the partial pressure of oxygen, reducing the drive force to sufficiently saturate the hemoglobin. Altitude changes from sea level to 10,000 feet reduce the oxygen levels, which significantly impacts conscious actions and can lead to incapacitation. This is only a reduction in atmospheric pressure by a third.

The process of evaluating the pulmonary system is called the pulmonary function test, which evaluates the resting breathing (tidal volume), the forced inspiration (inspiratory reserve volume), the forced exhalation (expiratory reserve volume), and the residual volume. From these measurements the vital capacity can be calculated, as well as the functional reserve capacity and the total lung capacity. Total lung volume ranges from 4.8 to 6.2 liters in males and 3.6 to 4.6 liters in females. The residual volume is the amount of air required to maintain lung inflation, which cannot be forced out. It ranges from 900 milliliters to about 1,300 milliliters, with an average of 1,200 milliliters. Tidal volume is a variable for each individual and the difference between the tidal volume and forced inspiration is the inspiratory reserve volume. The volumes derived from a pulmonary function test are used to diagnose chronic obstructive pulmonary disease, asthma, and others.

CHAPTER TEN

THE URINARY SYSTEM

ORGANIZATION AND FUNCTION OF THE KIDNEYS AND ASSOCIATED STRUCTURES IN MAINTENANCE OF FLUID AND ELECTROLYTE BALANCE

LEARNING OBJECTIVES

Upon completion of the chapter readers will have the essential knowledge of the urinary system's anatomy and physiology and the role of the urinary system in fluid and electrolyte balance:

1. Anatomy of the urinary system
2. Anatomy and function of nephrons
3. Importance of electrolytes and how they are balanced
4. Hormones involved in fluid and electrolyte balance
5. Impact of low oxygen tension and its relationship to the kidney

The urinary system consists of the kidneys (containing the functional unit of the *nephron*), the muscular *ureter* which drains the kidney, the *bladder* which temporarily stores the urine, and the *urethra* which empties the bladder. The system provides for the elimination of liquid metabolic waste, excess water, water-soluble vitamins, and excess minerals (particularly electrolytes); aids in the regulation of blood pressure; and promotes production of erythrocytes. The urinary system accomplishes these functions by filtering the plasma to separate wastes and recover the useful materials; regulating blood volume and pressure by conserving or eliminating of water; regulating the osmolarity of the blood by balancing the solute content of the plasma; secreting an enzyme *renin* to stimulate and activate *angiotensin II* to control blood pressure and maintain electrolyte balance; secreting the hormone *erythropoietin* to promote the production of erythrocytes and increase oxygen-carrying capacity; and functioning with the lungs to control the body's pH through the management of carbon dioxide content. Calcitriol is synthesized in the kidney and activated by the liver into vitamin D. The kidneys are involved in detoxification of free radicals and drugs, and *gluconeogenesis* from amino acids with the excretion of ammonia.

To regulate homoeostasis, the urinary system interacts with other systems to control pH, metabolic wastes, toxins, and water. This involves the respiratory system, where water in a vapor state and carbon dioxide are eliminated. The interaction with the integument involves the loss of water, electrolytes, and urea in sweat. The interaction with the digestive system eliminates water, salts, carbon dioxide, bile salts, cholesterol, and other metabolic wastes.

Nitrogenous Waste

Nitrogenous wastes are produced when proteins and amino acids are metabolized. To decompose an amino acid and reach the core carbon chain, the first step is the *deamination*, or the removal of the amino group, yielding ammonia, which is highly toxic to tissue. The ammonia is transported to the liver, which converts the ammonia to urea, a less toxic nitrogenous molecule and important in the maintenance of the osmolarity of the blood. The urea makes up about fifty percent of the nitrogenous wastes; uric acid and creatine are others that are produced by catabolic actions on nucleic acids and creatine phosphate. Collectively these nitrogenous wastes are the components measured in a blood sample tested for blood urea nitrogen (BUN) which ranges between ten and twenty milligrams per deciliter of blood. If the value exceeds the normal range it would indicate a condition of azotemia, which suggests kidney failure or insufficiency and may trigger diarrhea, vomiting, dyspnea, or cardiac arrhythmias.

Anatomy

The principle organ of the urinary system is the kidney, which is located along the dorsal abdominal wall between the levels of T12 and L3 vertebrae, with the right kidney being lower than the left due to the right lobe of the liver. The kidneys are attached posteriorly to the peritoneal membrane of the abdomen, so they are said to be retroperitoneal in location. The attached ureters, renal arteries, and veins, along with lymphatic vessels and adrenal glands, which rest along the superior medial margin of the kidney, are in a retroperitoneal location.

Kidney

Each kidney is a glandular organ composed of an estimated 1.2 million nephrons and each weighing around 160 grams. They average approximately 10 centimeters in length, 5 centimeters in width, and 2.5 centimeters in thickness; this is about the size of most bars of soap. The lateral side has a convex curve, while the medial side has a concave curve, creating the bean shape. In the medial side is a slit like the eye of a bean called the *hilum* where the ureter exits the kidney, and the blood, lymphatic vessels, and nerves enter the kidney. The kidney consists of three layers of connective tissue that protect the kidney and attach it to the abdominal wall. The outermost layer is the *renal fascia*, fibrous connective tissue binding the kidney to the parietal peritoneum and the abdominal wall. The middle layer is composed of an adipose capsule involved in cushioning the kidney and maintaining its location. The innermost layer is the *renal capsule*, a fibrous layer covering the outside of the organ, which is anchored in the hilum and protects the kidney from trauma and infection. Collagen fibers extend between the renal fascia and capsule, providing additional support to the kidney. Movement of the body from supine to vertical creates a change of position of the kidneys of about three centimeters. Any movement that shifts the kidneys more than three centimeters results in reduction of the blood flow or urine flow, creating a potential hazard to the kidneys' function.

The parenchyma (functional component) of the kidney is C-shaped, encircling the medial renal sinus. It is divided into two zones: an outer *renal cortex*, which is approximately one centimeter thick; and the inner *renal medulla*, facing the renal sinus. The renal cortex extends, creating *renal columns* through the medulla, dividing the medulla into six to ten segments or *renal pyramids*. The renal pyramids are conical shaped, with the base along the cortex and the peak forming the *renal papilla*, facing the renal sinus. The pyramid is composed of a series of papillary ducts that transport urine to the renal papilla. The renal papilla nestles in a cup called the *minor calyx*, where the urine is discharged and collected in the renal sinus. Several minor calyxes merge, forming the *major calyx*, all the major calyces merge into the *renal pelvis*. The renal pelvis acts as a funnel to direct the urine into the ureter to drain the urine to the urinary bladder by peristalsis and gravity.

Renal circulation

The renal fraction is about twenty-one percent of the cardiac output. The blood from the abdominal aorta is directed through the renal arteries, entering the hilum of the kidney and dividing into the segmental artery, routing the blood to the cortex through the interlobar arteries through the renal columns, arcuate arteries along the base of the renal pyramids, and then the interlobular arteries, then to the nephrons through the afferent arterioles. The capillary bed in the nephrons (glomerulus) is drained by the efferent arterioles and forms a secondary capillary bed around the distal convoluted tubules of the nephron called the *juxtaglomerular apparatus*, where peritubular capillaries collect electrolytes and water, and then drain into interlobular veins, then the arcuate veins, interlobar veins, segmental veins, and the renal veins, into the inferior vena cava. The juxtamedullary nephrons, which are located closer to the medulla, have a modified efferent arteriole pathway called the *vas recta*, which is a capillary bed surrounding the tubules that recovers more water from the filtrate.

Sympathetic nerve fibers regulate the blood flow to the kidneys, controlling the nephron filtration rate. Renin secretions are stimulated by the nervous activity, which is involved in controlling blood pressure.

Nephron

The functional unit of the kidney is the *nephron*, which consists of a renal corpuscle and a renal tubule. The corpuscle is constructed of a two-layer glomerular (Bowman's) capsule surrounding a capillary bed (glomerulus). These capillaries are *fenestrated*; that is, they have opening between the cells. The inner layer of the capsule has a series of cells called *podocytes* that wrap around the glomerulus and control the fluids and contents, leaving the glomerulus and entering the entering the capsular space. The fluid and content enter the proximal convoluted tubule as filtrate. The renal tubules are divided into regions, where functions vary to balance the fluid, electrolytes, and osmolarity of the blood. The first region attached to the capsule is the proximal convoluted tubule; the second is a loop called the *loop of Henle*, with thick and thin segments that ascend and descend, and the distal convoluted tubule that connects to the collecting duct. There are two distinct types of nephrons, based on their location. The *cortical nephrons* are located deep in the renal cortex and have short tubules. The *juxtaglomerular apparatus* and the *juxtamedullary nephrons*, which are located near the medulla and have longer loops, are associated with the *vas recta*. The cortical nephrons are the most abundant, comprising eighty-five percent of the nephrons. The juxtamedullary nephrons are essential for water conservation in the body.

Urine formation

Urine is a product of the filtered wastes and water from the blood plasma. The formation of urine begins with fluids from the blood plasma, with included electrolytes and organic molecules that enter the capsular space, becoming the glomerular filtrate. This filtrate contains very little protein, most of which are very small protein molecules, smaller than albumin. From the capsular space the filtrate moves into the tubules, becoming tubular fluid, with its composition being modified as water is recovered and exchanged with the surrounding tissue or selectively secreted back into the filtrate. At the end of the tubule the filtrate becomes urine and enters the collecting duct, where slight modifications to its composition occur until entering the papillary duct.

In the glomerulus there is a filtration membrane that has three barriers controlling fluid passage. The first barrier is the fenestrated endothelium of the glomerular capillaries, which prevents formed elements from exiting the plasma. The second barrier is the basement membrane, which has a negative membrane charge and pore sizes that exclude proteins like albumin. The third barrier is the filtration slits covered by the negatively charged podocytes, which exclude anions from the filtrate. Glomerular filtration is driven by the blood hydrostatic pressure, which is approximately 60 mmHg; the capsular hydrostatic pressure, which is approximately 18 mmHg; and a colloidal osmotic pressure of approximately 32 mmHg in the blood. The glomerular filtrate lacks components necessary for a colloidal osmotic pressure, creating a pressure gradient between the filtrate and the blood. The net filtration pressure is 10 mmHg out of the blood driving the filtrate. This pressure enables the production of approximately 12.5 milliliters per minute of filtrate per mmHg of filtration pressure, or about 125 milliliters of filtrate per minute (180 liters per day for men and 150 liters per day for women). The majority of the filtered fluid is recovered in the tubules, with the daily urine output between 1 to 2 liters each day.

The glomerular filtration is controlled by renal autoregulation, which is a myogenic mechanism related to the stretch response of smooth muscles in the afferent arterioles and the *tubuloglomerular feedback,* monitoring the distal convoluted tubule filtrate and triggering the release of rennin by the juxtaglomerular cells. Dilation and constriction of afferent and efferent arterioles are controlled by the *macula densa* under sympathetic nervous control. The rennin-angiotensin mechanism stimulates vasoconstriction throughout the body. The constriction of the afferent and efferent arterioles reduces water loss and lowers the filtration rate. It stimulates the reabsorption of water and sodium chloride (NaCl) in the proximal convoluted tubule. Additionally, the adrenal cortex is stimulated to secrete *aldosterone*, which promotes sodium and water retention. The secretion of *antidiuretic hormone* (ADH) by the posterior pituitary gland stimulates thirst and water reabsorption.

The proximal convoluted tubules reabsorb about sixty-five percent of the original filtrate by transcellular or paracellular mechanisms. Sodium is reabsorbed, along with chloride and water. Any bicarbonate ions in the filtrate are removed by the formation of bicarbonates in the blood. Electrolytes (potassium, magnesium, and phosphates) diffuse by paracellular routes along with the water. Glucose in the filtrate is cotransported with the sodium. Between forty and sixty percent of the urea is reabsorbed with water; the uric acid reabsorbed in the proximal tubule will be excreted later in the tubule. The blood originally entering the kidney contains approximately 20 milligrams of urea per deciliter, while in the blood in the renal veins the urea has been reduced to approximately 10.4 milligrams per deciliter. The cells along the tubules will absorb small peptides. Small concentrations of bile acids, ammonia, catecholamine, prostaglandins, and creatinine are secreted as well as hydrogen ions in the distal tubules.

The nephron loop generates a salinity gradient that concentrates the filtrate by further absorption of water from the filtrate. Approximately twenty-five percent of the sodium, potassium, and chlorides are absorbed with fifteen percent of the water in the filtrate. The potassium is then secreted back into the filtrate.

The distal convoluted tubule reabsorbs water and additional salts. This reabsorption is regulated by aldosterone, promoting the reabsorption of sodium in the thick segment of the loop, the distal convoluted tubule, and cortical section of the collecting duct. *Aldosterone* promotes the secretion of potassium while the chlorides and water accompany the sodium reabsorption. *Atrial natriuretic peptide* (ANP) that is produced in the atrial myocardium promotes the secretion of sodium and water, reducing the blood volume and increasing the urinary output by dilating the afferent arterioles and constricting the efferent arterioles, increasing the filtration rate, which antagonizes the angiotensin-aldosterone mechanisms by inhibiting the effects of the enzyme rennin and the secretion of aldosterone by the adrenal gland. *Antidiuretic hormone* (ADH) secretion is inhibited as well as its actions in the tubules. These actions collectively inhibit the recovery of sodium and chlorides from the filtrate in the tubules and the urine in the collecting ducts. ADH is released in response to lowering blood pressure associated with dehydration. The ADH raises blood osmolarity and makes the ducts more permeable to water, increasing the recovery of water from the filtrate, concentrating the urine. Parathyroid hormone stimulates excretion of phosphates in the proximal convoluted tubules and promotes the recovery of calcium in the thick loop and distal convoluted tubule along with stimulating calcitriol synthesis in the proximal convoluted tubule.

The urine produced by the actions of the nephrons, tubules, collecting and papillary ducts is composed mostly of water and associated minerals and wastes. It collects in the renal pelvis and enters the ureter, which moves the urine by peristalsis and gravity to the bladder for storage. The ureter is a tube composed of three layers: the outer layer, called the adventitia; the middle layer, composed of smooth muscle; and the inner mucosa lining. The urine is stored in the urinary bladder, which is capable of containing approximately 700 to 800 milliliters. The bladder is a muscular sack located on the pelvic floor, posterior to the pubic symphysis. It is covered by the parietal peritoneum and a fibrous adventitia. The *muscularis* consists of the *detrusor muscle*, which is three layers of smooth muscle; and a mucosa, which responds to stretching and is wrinkled in the relaxed state. The ureter entry and the urethra exit are located in a triangular area called the *trigone*. The urethra conducts the urine to the exterior. It is approximately eighteen centimeters in length in males, divided into three regions: the prostatic region, the membranous region, and the penile region or urethra. In females the urethra is approximately three to four centimeters long. There are two sphincter muscles that control the retention of urine in the bladder; the internal urethral sphincter is smooth muscle, and the external urethral sphincter is composed of skeletal muscle. Voiding of urine is called *micturition* and is controlled by the stretch feedback to the pons and by the external sphincter, which individuals learn to control between the ages of two and three years.

Urine has an appearance of a clear liquid ranging from colorless to pale yellow, darker straw, or dark amber, depending on the components other than water contained in the urine. The normal straw yellow color is contributed by urochrome, which is formed by bacterial action on bilirubin. Other colors are due to food, drugs, and breakdown products. The odor of urine is slightly astringent but can range widely depending on the foods and medications the individual is consuming. Normal specific gravity of urine range is between 1.000 and 1.028, depending on the hydration level of the individual. The pH ranges from 4.5 to 8.2, but most will be between 6 and 7. The urine is normally ninety-five percent water and five percent solutes. The most abundant organic solute is urea. The most abundant inorganic solute is sodium chloride, with lesser amounts of potassium, uric acid, creatine, phosphates, sulfates, calcium, magnesium, and urochrome. Most individuals eliminate around 1.5 to 2 liters of urine daily.

Fluid, Electrolyte, and Acid-Base Balance

The primary fluid in the body is water. An adult's body consists of between fifty-five and sixty percent water, while at birth the body is about seventy-five percent water. Water is distributed into compartments, with sixty-five percent of the fluid contained in the intracellular fluids (inside the cells) and thirty-five percent in the extracellular spaces, which include the interstitial fluid (around the tissues), blood plasma, lymph, and transcellular fluids. The fluids are in constant motion, moving from one compartment to the next by osmosis. The body loses approximately 2,500 milliliters of water per day through all mechanisms. Some water is consumed in foods, which is preformed water; other water is produced metabolically as energy is produced. About 1,500 milliliters of water is excreted each day as urine, 200 milliliters is lost in feces, 300 milliliters is lost in respiration, 100 milliliters in sweat, and 400 milliliters through transpiration by the skin. The activity level will vary the amount of water lost daily through all sources. The fluid intake should balance the fluid loss to maintain the blood volume and osmolarity.

Sodium is the electrolyte responsible for the resting membrane potentials, and triggers the depolarization. It accounts for ninety to ninety-five percent of the osmolarity and is the principal cation found in the extracellular fluid compartment (ECF). Sodium molecules are abundant, with sodium bicarbonate being the principal molecule involved in the buffering of the pH in the small intestine. The sodium potassium pumps generate heat for the body. Daily need for sodium is 0.5 grams, with the typical diet having 3 to 7 grams. Sodium levels are managed by the interactions of several hormones. Aldosterone promotes sodium recovery and retention in the body. Antidiuretic hormone is a water conserver, but recovers sodium passively along with the water. Atrial natriuretic peptide inhibits sodium recovery from the filtrate and increases water output. Excess sodium in the body promotes water retention, hypertension, and edema. Insufficient sodium is not normally problem because the body adjusts the urinary output to maintain the sodium balance.

Potassium is the cation that interacts with sodium, creating the membrane resting potentials in nerves and muscles. The intracellular fluid compartment (ICF) has a higher concentration of potassium, which is an essential cofactor for protein synthesis. The concentration of potassium is managed by aldosterone, which stimulates it secretion in the distal tubules. Excess potassium creates an abnormally excitable state in the nerve and muscle, which can result in cardiac arrest. Elevated potassium slows the heart rate. Low concentrations of potassium can result muscle weakness, depression of reflexes, and irregular cardiac muscle activity (frequently uncontrolled tachycardia). Low levels of potassium can result from heavy sweating, vomiting, diarrhea, and some blood pressure medications.

Calcium is the most abundant cation in the body, with the majority bound in the skeleton. This electrolyte is involved in muscle contraction, nerve transmission, and clotting. The serum concentration of calcium is regulated by the interactions of parathyroid hormone, calcitriol, and calcitonin. Excesses calcium results in alkalosis. The membrane permeability of sodium is disrupted in nerve and muscle cell depolarization, creating weakness and cardiac arrhythmia. Low calcium levels make the nerves and muscles hyper excitable.

Chloride is the most abundant anion in the ECF and with sodium is a major contributor to osmolarity in the body. It is involved in the regulation of pH and vital to the production of hydrochloric acid in the stomach. Its balance is associated with sodium; it is transported along with water and sodium in the tubules. Acid-base difficulties can occur when chloride levels are unbalanced.

Phosphates are anions most abundant in the ICF. Phosphates are involved in energy transfer, intercellular buffering systems, and intercellular stimulation. Plasma levels are controlled by the interactions of parathyroid hormone and calcitonin.

Acid-base balance is maintained by the interaction of numerous buffer systems. The pH range is maintained between 7.35 and 7.45. The primary physiological buffer system in the ECF is the carbonic acid–bicarbonate system in the plasma, which involves the respiratory system's removal of carbon dioxide and the urinary removal of hydrogen ions. The buffer system absorbs and releases hydrogen ions to stabilize the pH range. The primary buffer system in the ICF is a phosphate buffer system, along with protein buffers.

The respiratory system adjusts the pH more rapidly with the dissociation of the carbonic acid into carbon dioxide, which is expelled by the lungs and leaves water in the blood. The urinary system is more effective than the respiratory in the excretion of hydrogen ions by the distal tubules and collecting ducts, which leaves the bicarbonate ions in the blood to rapidly bind with free hydrogen ions, reducing the pH.

LABELING ACTIVITY

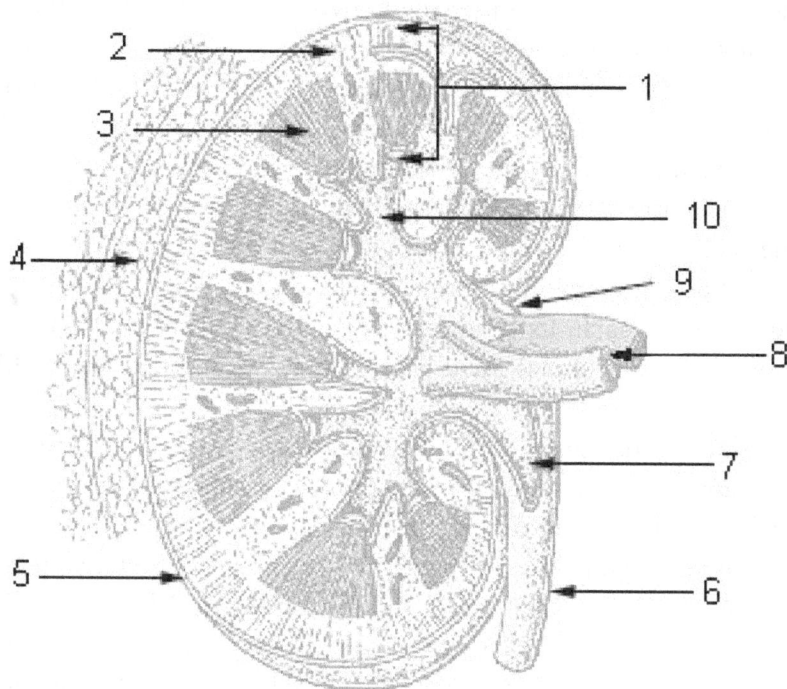

Fig. 10.1 Source: https://commons.wikimedia.org/wiki/File:Illu_kidney.jpg.

CHAPTER ELEVEN

THE DIGESTIVE SYSTEM

ORGANIZATION AND FUNCTION IN OBTAINING ENERGY AND NUTRITION REQUIREMENTS

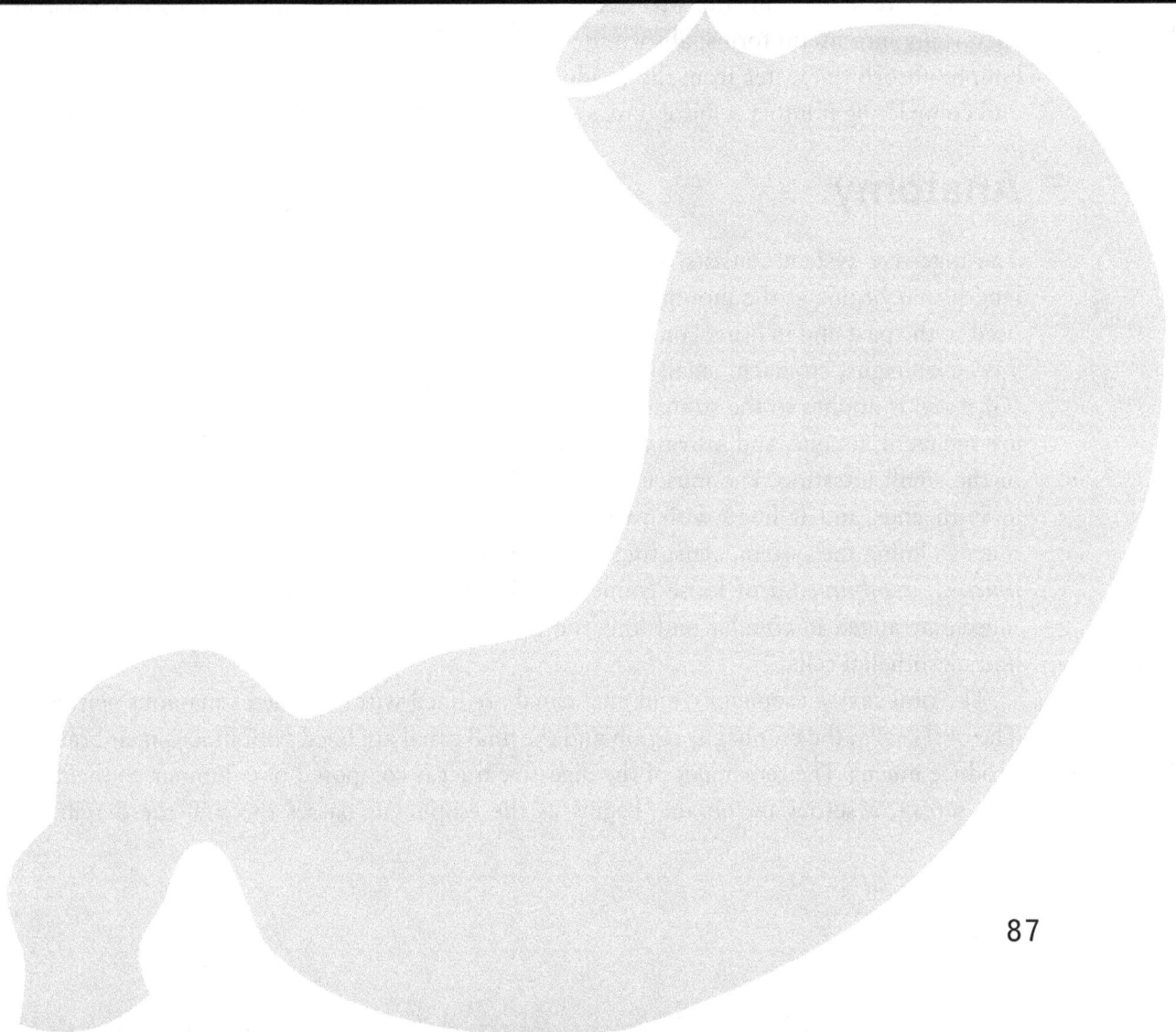

LEARNING OBJECTIVES

Upon completion of the chapter readers will have the essential and basic knowledge of the digestive system, its physiology, and its anatomy, including the accessory components required to maintain nutrient homeostasis:

1. Anatomy of the digestive system
2. Chemical and mechanical processes of digestion
3. Description of the flow of the digestive processes
4. Neurological and physiological control of the digestive system including hormonal management of the processes
5. Basic nutrition and how the organic groups are utilized
6. Accessory organs associated with the digestive system and their basic function

The digestive system functions to selectively ingest foods, digest those foods mechanically and chemically into useful forms; absorb the nutrients liberated during digestion into the blood and lymph; absorb the water from the residual mass of nondigested components, cells, and mucus, and compacting it into a semisolid mass (feces); and eliminating the mass by defecation.

Anatomy

The digestive system consists of a muscular tract approximately thirty feet (nine meters) in length that begins at the mouth and extends to the anus. Alimentary canal has been the term used in the past and in other contexts. The digestive system is composed of the mouth, oropharynx, esophagus, stomach, small intestine, and large intestine. The term gastrointestinal tract (GI tract) is applies to the stomach and intestines. Accessory structures of the digestive system are the teeth, tongue, and salivary glands in the mouth; and the liver, gallbladder, and pancreas in the small intestine. The muscular tube of the digestive system is open to the environment at both ends and is lined with various epithelial tissues. The basic plan of the system is a mucosa lining the system, consisting of epithelial cells, a *lamina propria*, and a thin *muscularis mucosa*; a *submucosa* of loose connective tissue; a *muscularis externa* composed of smooth muscle arranged in circular and longitudinal layers; and a *serosa* composed of areolar tissue and mesothelial cells.

The oral cavity, esophagus, and anal canal are lined with stratified squamous epithelial cells. The oral cavity, the esophagus region and the anal canal are lined with mucus membranes which produce mucus. The remainder of the digestive tract is composed of columnar epithelial tissues. The serosa, a serous membrane, begins as the esophagus passes through the diaphragm and extends to the rectum. The upper esophagus and the rectum are surrounded by fibrous connective tissue called the adventitia, which connects them to other structures in the area.

When the esophagus passes through the diaphragm and enters the abdomen, a series of sheets of connective tissue (mesenteries) support the structures, called the viscera. The mesenteries are formed from the parietal peritoneum as folds. These folds encompass the structures and support the blood, lymphatic vessels, and nerves. These mesenteries are the dorsal mesentery, ventral mesentery, *lesser omentum*, *greater omentum*, and the *mesocolon*. The abdominal viscera are considered to be intraperitoneal with the exception of the duodenum, pancreas, and sections of the large intestine, which are retroperitoneal, located between the peritoneum and the abdominal muscular wall.

Mouth

The mouth (oral or buccal cavity) functions to ingesting food, triggering sensory response to the food, initiating the mechanical digestion of the food by chewing (mastication), initiating the chemical digestion of food by mixing the enzymes in the saliva with the food, and ventilating the lungs and forming of sounds (speech). The oral cavity extends from the opening between the lips (oral fissure) to the *fauces* (posterior margin of the soft palate). The cavity is lined with squamous epithelial cells; those between lips and gums in the vestibule are keratinized, and *labial frenulum* attaches the lips to the gums. Portions of the cheeks are stratified squamous epithelial cells in the mucus membrane. The tongue, which is muscular, agile, and sensitive, occupies the inferior portion of the cavity and is attached by the lingual frenulum to the muscular floor. The upper surface of the tongue is covered by *lingual papillae*, where the taste buds (chemo receptors) for salt, sweet, sour, and bitter are located. The fifth taste, *umami*, which is described as the savory flavor associated with meat or proteins, is suggested to be at the base of the tongue near the oropharynx. The muscles of the tongue are a combination of intrinsic (within the tongue) muscles and the extrinsic (outside of the tongue) muscles. Along the base of the tongue are located the lingual tonsils, which are lymphatic structures, and other lingual glands. The superior portion of the cavity consists of the hard and soft palates. The uvula is an extension of the soft palate that contains nerves and the center of the gag reflex. Along the lateral wall behind the uvula are located the palatine tonsils, which are lymphatic structures. Located laterally in the upper and lower jaws are the teeth (dentition), which serve to masticate (chew) food to begin mechanical digestion. The teeth differ in shape and function required to chew the food. The incisors are used to cut or shear food; there are two on each side of the midline. The canine teeth function to puncture, tear, or shred the food; there is only one on each side of the midline. The premolars (bicuspids) are involved in crushing food, increasing its surface area; adults have two on either side of the midline and none in the deciduous set (baby teeth). The molars are involved in grinding food; there are three in the adult set and two in the deciduous set. The teeth are attached to the underlying bone (maxilla or mandible) by the *periodontal ligament*, in sockets formed in the alveolus in a gomphosis articulation. The portion of the tooth that extends into the oral cavity (crown) is covered with the acellular enamel. The portion of the tooth that extends below the gum (*gingiva*) (root) is composed of dentin and cementum. The majority of the mass of the tooth is composed of *dentin*, while the *cementum* is mostly located in the root. These are the living connective tissues of the tooth. The pulp cavity is an open area in the crown containing blood and lymphatic vessels, nerves, and loose connective tissue, which enter the cavity through canals in the root of the tooth.

The act of chewing (*mastication*) reduces the size of the food ingested, increasing the surface area for the enzymes to initiate chemical digestion. It mixes the saliva produced by the parotid, submandibular, and sublingual salivary glands with the food. The saliva is ninety-seven percent water. It contains *amylase*, an enzyme that breaks down plant starch into maltose; *lingual lipase*, which is activated in the acid environment of the stomach to break fats into fatty acids; *lysozymes* to inactivate bacteria by damaging the cell membranes; *immunoglobulin A* to inhibit bacterial growth; mucus to lubricate the food particles to ease swallowing and peristalsis in the esophagus; and electrolytes. The pH of the mouth and saliva is between 6.8 and 7.0. The food is formed into small masses called *bolus* that are swallowed by the constriction of the pharynx (throat) muscles where they enter the esophagus, which is twenty-five to thirty centimeters (ten to twelve inches) in length, moving the bolus by peristalsis to the stomach, where it empties into the stomach through the *cardiac valve* (orifice).

Stomach

The stomach is a J-shaped organ with an inner (lesser) curvature and an outer (greater) curvature. There are four regions to the stomach: the location where the esophagus empties is the cardiac region; the domed area extending above the esophagus entry point is called the *fundus* (fundic) region; the lower funnel-shaped area is the *antrum*; and the *pyloric region* (canal) is where the stomach contents empty through the *pyloric sphincter* into the duodenum of the small intestine. Three layers of smooth muscle make up the muscularis: the inner muscle layer is arranged obliquely; the middle layer is arranged circularly; and the outer layer is arranged longitudinally. Internally, the stomach is lined with a *ridged mucosa* (*rugae*) with a series of gastric pits containing *parietal cells*, which produce the hydrochloric acid and intrinsic factor; and chief cells, which produce *pepsinogen* and *gastric lipase*. In the lining of the cardiac and pyloric regions there are enteroendocrine cells, which produce local regulatory hormone for the digestive system. Each day the stomach secretes two to three liters of gastric juices that mix with the ingested food. These gastric juices contain water, hydrochloric acid, and pepsin (activated pepsinogen). The hydrochloric acid disrupts connective tissue and cell walls; activates the pepsinogen into pepsin; converts ferric (iron 3) to ferrous (iron 2), which is absorbed and used for hemoglobin synthesis; and destroys most ingested microorganisms. The pepsin initiates the protein breakdown into shorter peptide chains. The gastric and lingual lipase liberate fatty acids from fifteen percent of the dietary fats in the foods. The *intrinsic factor* is essential to promote the B12 absorption vital to hemoglobin production. The stomach churns the foods with the gastric juices, creating a fluid *chyme*, along with the separation of the carbohydrates, proteins, and fats from the ingested food, with a pH between 1.8 and 2.2. Approximately two hours after the ingestion of the food, the carbohydrates are passed through the pyloric sphincter into the duodenum; four hours after ingestion, the proteins are ejected into the duodenum; and after six hours the fats have finally passed from the stomach into the duodenum. Each chyme is released in approximately three milliliter batches into the duodenum. The only nutrients absorbed by the stomach are water soluble, particularly ethanol and aspirin. The enteroendocrine cells produce chemical messengers (local hormones) that promote emptying of the stomach, secretion of gastric juices, inhibition of gastric juices, and the churning of the stomach. These chemical messengers include substance P, vasoactive intestinal peptide, secretin, gastric inhibitory peptide, cholecystokinin, and neuropeptide Y. They regulate and interact with the stomach, intestine, and nervous system to control various aspects of the digestive process. The stomach activity is divided into three stages: the *cephalic stage* is in the brain, where smelling food, seeing food, thinking about food, and tasting food stimulates the vagus nerve root to initiate the preparation of the stomach to receive food; the *gastric phase* is activated by swallowing food to stimulate the gastric secretions, which are stimulated by histamines, acetylcholine, and gastrin; and the *intestinal phase*, stimulated when chyme is ejected into the duodenum, enhancing gastric secretions; and with the arrival of semidigested fats the enterogastric reflex is initiated, inhibiting secretions and churning of the stomach through the actions of secretin and cholecystokinin, which also initiate the pancreas and gallbladder release of bile and pancreatic juices.

Small Intestine

The small intestine is divided into three segments. Total length is 2.7 to 4.5 meters in length, with a diameter of about 2.5 centimeters. The first segment is the *duodenum*, which is about 25 centimeters in length, beginning at the pyloric valve and becoming the *jejunum* at the duodenojejunal flexure; the pancreatic juices and bile are released into the duodenum, adjusting the pH of the chyme to about 7.4 to 7.8. The jejunum makes up about forty percent of the small intestine and is 1 to 1.7 meters in length. A characteristic of the jejunum is a rich blood supply for absorption of nutrients and a series of enzymes to promote further chemical digestion. The last segment is the *ileum*, which is about sixty percent of the small intestine, being 1.6 to 2.7 meters in length. The ileum has a well-developed series of lymphatic structures organized into the *Peyer patches*. The ileum ends at the ileocecal valve. Throughout the small intestine villi, folds called *plicae circulares*, increase the surface area for chemical digestion and absorption of nutrients to approximately 200 square meters. Embedded in the *lamina propria* are *lacteals*, which are involved in the absorption and transport of fats. The villi are the location of the chemical digestion by the brush border enzymes. Intestinal secretions add one to two liters to the volume daily in response to the acid, hypertonic chyme, and distension. Motility of the small intestine creates segmentation of the nutrients in the process of moving the chyme to the colon.

Large Intestine

The large intestine begins at the ileocecal valve and extends to the anus. The *ileocecal valve* empties into the *cecum*, a blind pouch, with the attached *vermiform appendix*, found to be important in lymphatic system. The entire large intestine is referred to as the colon. The cecum is attached to the ascending colon, extending along the right side of the abdomen until it turns left at the hepatic (right colic) flexure, becoming the transverse colon across the upper abdomen. The transverse colon turns inferiorly at the splenic (left colic) flexure, becoming the descending colon and changing into the lazy S-shaped sigmoid colon joined to the rectum. The exterior of the large intestine has a very obvious *taenia coli*, a longitudinal band of smooth muscle; and obvious pouches called *haustra*, where bacterial activity converts residue into vitamins, and water is absorbed. At the anus an internal sphincter composed of smooth muscle, and an external sphincter composed of skeletal muscle, regulate the elimination of the solid wastes. The water absorption recovers vitamins and electrolytes released by microbial activity. The peristalsis and the thickened residue is compacted for elimination as semisolid mass containing residue of the foods that could not be digested, dead bacteria, and cells, mucus, and other ingested material.

Digestive System Accessory Structures

The *liver* is the largest organ, weighing approximately three pounds. It is divided into a large right lobe that impinges on the right lung. The right lobe is connected to the *falciform ligament*, to the quadrate lobe next to the gallbladder, and the tapered caudate lobe. The caudate lobe has an irregular opening where the hepatic portal enters, called *porta hepatic*.

The *gallbladder* is a pear-shaped bladder located near quadrate lobe that stores bile and excess cholesterol. The fluid is concentrated, leading to the development of calculi (stones). The bile and cholesterol are transported from the liver to the gallbladder by the bile duct. The gallbladder releases bile containing cholesterol and bile salts to the pancreas through the cystic duct.

The *pancreas* is a mixed organ, producing endocrine and exocrine chemicals. It produces about 1.2 to 1.5 liters of pancreatic fluid a day, which is released into the duodenum. This juice is produced by acini, which are connected to the pancreatic duct, and delivered along with bile. Sodium bicarbonate in the pancreatic juice adjusts the pH of the chyme received from the stomach. The juice contains a diverse and abundant series of enzymes: zymogens, which convert into trypsin; enterokinase; chymotrypsin; amylase; lipase; and nucleases. The release of the pancreatic juice and bile is stimulated by the interactions of acetylcholine, cholecystokinin, and secretin.

Digestive States

The consumption of food and appetite is regulated by a group of peptide hormones acting in the short or long term. Short-term regulator ghrelin is a local hormone secreted by the fundus of the stomach during the hour after eating, and ceases secretion at that time. Peptide YY (PYY) is a neuropeptide secreted by the ileum before the chyme reaches it, signaling satiety (fullness) to terminate the ingestion and slow the emptying of the stomach. Cholecystokinin (CCK) is secreted by the duodenum and jejunum, which stimulate pancreatic juice and bile release and effect the brain and sensory fibers of the vagus nerve, stopping ingestion of food. The long-term regulators control appetite, metabolic rates, and body weight over longer periods of time. Leptin is a regulator secreted by adipocytes throughout the body, regulating hunger and ingestion; a deficiency leads to overeating and is used medically to control obesity. Insulin produced by the pancreas beta cells stimulates glucose and amino acid uptake by muscles and other tissues, and promotes the synthesis of glycogen and fats. In the brain the arcuate nucleus regulates appetite by secreting neuropeptide Y (NPY) to stimulate appetite and melanocortin to suppress appetite. The sense of hunger stimulates gastric peristalsis when the stomach is empty during the cephalic phase. The feeling of satiation is generated by the distension of the stomach, but is short term without absorption of nutrients into the blood. Cravings for specific types of food are driven by neurotransmitters. Norepinephrine stimulates a craving for carbohydrates, galanin a craving for fats, and endorphins a craving for proteins.

Following ingestion, an absorptive state lasting about four hours takes place, with an increased glucose level in the blood. Excess glucose is stored in muscles and liver as glycogen or converted by adipocytes into fat. The absorptive state is regulated by insulin promotion of glucose uptake and oxidation that inhibits gluconeogenesis. After about four hours the postabsorptive state develops, regulated by glucagon produced in the pancreas; epinephrine and norepinephrine promote lipolysis and glycogenolysis. This is the state between meal and overnight, when the body is using stored fuels. Cortisols promote the catabolism of fats and proteins, glucagon promotes the gluconeogenesis, and growth hormones raise the blood glucose by antagonizing the actions of insulin.

The body's metabolic rate varies due to physical, mental, and hormonal states. The basal metabolic rate (BMR) is a reference based on a comfortable, resting awake, postabsorptive state, which is about 2,000 kilocalories per day; light physical activity increases the rate to 2,500 kilocalories per day, and heavy activity may increase it to 5,000 kilocalories per day or more. This varies with age, gender, mental state, stress, and physical state. Younger growing individuals' BMR are not the same as adults; this is very true for individuals in puberty.

Nutrition

Nutrition is providing needed minerals, organic molecules, and energy sources to the body. Typically, nutrition is measured in the number of calories ingested each day. The calorie used in nutrition is actually one thousand times greater than that used in chemistry and physics to measure potential energy. The food ingested contains potential energy to support physiological activity. Most of the energy is derived from carbohydrates, proteins, and fats. Most proteins contain 4 kilocalories per gram and fats contain 9 kilocalories per gram. Alcohol, with 7.1 kilocalories per gram, and high-sugar foods rapidly spike the blood glucose and are considered empty calories, which could lead to malnutrition by suppression of the appetite.

Nutrients in the body are divided into macro and micro nutrients. Those used for energy—carbohydrates, proteins, fats, and water—are macro nutrients. The nutrients needed in small concentrations are micro nutrients, mostly minerals and vitamins. There are eight amino acids that cannot be made in the body and so must be obtained from food. These are essential amino acids. Daily dietary needs are 2.5 liters of water; 125 to 175 grams of carbohydrates; 80 to 100 grams of lipids; 44 to 60 grams of proteins; 0.05 to 3,300 milligrams of minerals and 0.002 to 60 milligrams of vitamins. Various nutrients have different benefits and deficits when used to provide energy to support activities.

Carbohydrates

Carbohydrates are the easiest molecules metabolized to produce energy, with the fewest risks. To fully metabolize carbohydrates adequate oxygen is required to harvest all of the available energy in the molecules. Since the common carbohydrates available are polymers of glucose or other hexose sugars that can be converted to glucose, glucose is the model used for carbohydrate metabolism. Coenzymes nicotinamide adenine dinucleotide (NAD) and flavin adenine dinucleotide (FAD) are required to capture electrons and hydrogen ions for transfer between the metabolic pathways. Complete oxidation of glucose begins with the glycolysis occurring in the cytoplasm, with the enzymatic splitting of the central carbon-to-carbon bond to produce two pyruvic acid molecules and four adenosine triphosphate (ATP) molecules. This only yields a net of two ATP, since two ATP are required, along with the enzymes, to break the central carbon-to-carbon bond. In the absence of available free oxygen, the pyruvic acid is reduced to lactic acid, which regenerates the NAD required to sustain glycolysis, producing additional ATP. In the presence of free oxygen, the pyruvic acid in the presence of coenzyme A loses a carbon as carbon dioxide, and generates NADH plus an additional hydrogen ion, becoming an acetyl group attached to the coenzyme A. The Acetyl-Coenzyme A complex enters the tricarboxylic cycle (TCA or Krebs) located in the mitochondria. In the presence of the coenzyme A, the acetyl radical is joined to a four-carbon molecule, creating citric acid and releasing the coenzyme A. The cycle continues with the loss of carbons as carbon dioxide; hydrogen ions and electrons to NAD; and hydrogen ions to FAD. Coenzyme A re-enters the cycle, creating conditions where a high-energy phosphate is attached to guanosine diphosphate (GDP), creating GTP. The aerobic pathway of the Krebs cycle produces six NADH plus six hydrogen ions, two $FADH_2$ and two GTP, which yields thirty or thirty-two ATP molecules. The actual production of the ATP occurs in the electron transport chain by a series of electron transfers, and proton pumps creating water and producing ATP by ATP synthase. Aerobically, this occurs without developing lactic acid, which becomes toxic in the tissue when it builds up excessively. This process varies slightly tissue to tissue. Glucose can by synthesized from glycerol derived from fats or from amino acids by gluconeogenesis reaction.

Lipids

The majority of the lipids that are metabolized are derived from triglycerides stored in adipocytes. The triglycerides are composed of three fatty acid chains attached to a central glycerol. They may have been synthesized from various dietary sources by lipogenesis. To metabolize a fat it must undergo lipolysis, which is hydrolyzing the molecule, releasing fatty acids in sequence and eventually glycerol. Each fatty acid chain undergoes beta oxidation in the mitochondrial matrix, cleaving a two-carbon acetyl group, which joins with coenzyme A to enter the Krebs cycle, producing approximately 129 ATP molecules per fat molecule when completely oxidized. When fat oxidation is incomplete, ketones are created by ketogenesis. Some tissues use ketones to produce energy, but an excess of ketones can create an acidotic state called ketoacidosis, which is very dangerous and can occur in drastic weigh loss conditions with certain diets.

Proteins

The turnover of proteins averages about a hundred grams per day and is highest in the intestinal mucosa. Free amino acids are used to synthesize new proteins, converted to glucose and fats for storage or directly oxidized for energy. The catabolism of amino acids requires the removal of the amino group (deamination), which produces the toxic ammonia, which combines carbon dioxide with two ammonias to produce urea. The remaining carbon chain is converted to an acetyl radical, which is used to produce ATP.

Thermoregulation

Thermoregulation in the body maintains the body's temperature. The core temperature of the body estimated by rectal temperature is between 37.2 and 37.6 degrees Centigrade, with an outer temperature taken orally ranging between 36.6 and 37 degrees Centigrade. Heat in the body is generated by the exergonic reactions in the brain, heart, liver, skeletal muscles, and endocrine glands. The greatest source of heat in a resting individual is the brain and liver. Heat loss occurs by radiation, conduction, and evaporation. The hypothalamus monitors the temperature of the blood and thermal receptors in the skin, stimulating vasodilation or vasoconstriction and shivering to maintain the temperature. Vasodilation initiates sweating; vasoconstriction stimulates the removal of blood from the skin, the skeletal muscles to rapidly contract and relax, and smooth muscles to contract, creating a dead air insulation on the skin's surface. An increase of the metabolism creates heat in the body, signaling a response of reducing body coverage. Excessive heat in the environment or the body can induce problems. Heat cramps, heat exhaustion, and heat stroke create physiological challenges for the body. Without intervention these may lead to multi-organ dysfunctions that, if uncorrected, can lead to death. Hypothermia occurs when the core temperature reduced below 37 degrees Centigrade, which with environmental temperatures below 32 degrees Centigrade can be fatal by continuous heat loss thickening the blood, increasing the cardiac workload.

CHAPTER TWELVE

THE REPRODUCTIVE SYSTEM

ORGANIZATION AND FUNCTION OF HUMAN STRUCTURES INVOLVED IN PRODUCING NEW INDIVIDUALS

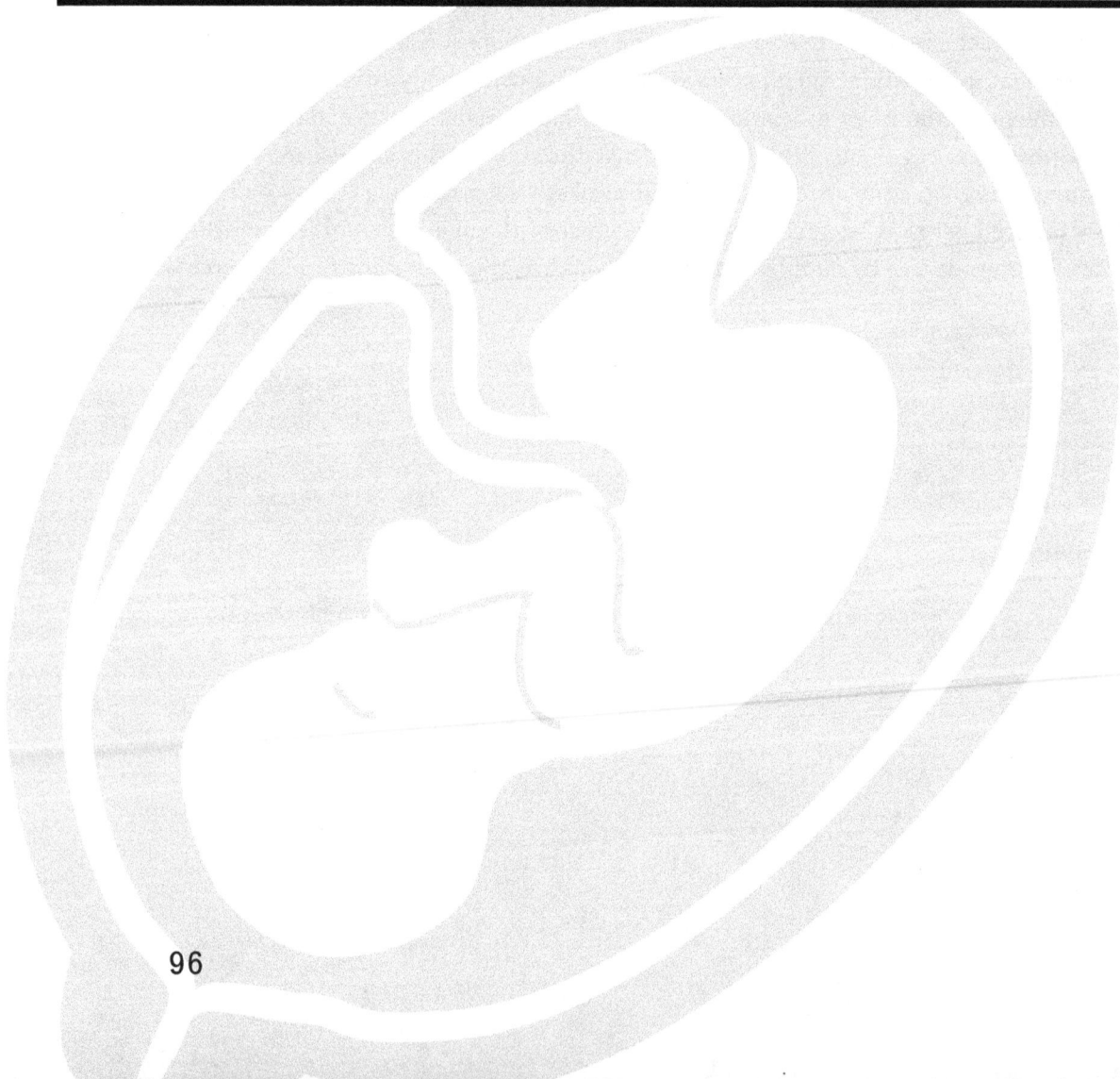

LEARNING OBJECTIVES

Upon completion of the chapter readers will have essential knowledge of the male and female reproductive systems, hormonal influences, and the gestational processes:

1. Anatomy of the male and female systems and their embryology
2. Roles of the various hormones in the development and maturation of the reproductive system
3. Basic physiology of male and female reproductive processes
4. Basics of conception and hormonal responses involved in implantation
5. Hormonal changes during pregnancy and delivery

Human reproduction is the production of a new individual by the combination of genetic materials (genes) produced by two other individuals (parents), with one contributing genetic material only (sperm) and the other contributing genetic material plus other cellular components (ova). The union of the sperm and ova creates a fertilized egg called a zygote, which with successful development produces a new individual. The sperm produced by the male is a small motile packet of genetic material; the ova produced by the female is a large, immobile, nutrient-rich cell possessing the genetic material. When fertilized, it develops over a period of approximately nine months into a baby. From this baby develops the next generation. For the production of gametes, the individual must mature and develop secondary and primary sexual characteristics and mature sex organs from the original genetic code at conception. The male develops the glands, ducts, and penis to a stage where sperm is produced and anatomical development is adequate for the delivery of the sperm into a receptive female. In the female, the fallopian tubes, uterus, and vagina must develop to accept the male; and the glands must develop to stimulate the ovaries to continue the oogenic possesses and the release of an ova. The external genitalia are located in the perineal area, while the internal structures are located in the pelvic cavity. Human genetics consist of twenty-three pairs of chromosomes; twenty-two pairs of autosomes, which determine many anatomical and functional characteristics, and one pair of chromosomes, which determines the sex and sexual characteristics of the individual. The sex chromosomes consist of a large X chromosome and a smaller Y chromosome. An individual who is genetically female possesses two large X chromosomes, while a male has a large X and a smaller Y chromosome. The Y chromosome carries the coding for the production of *testis determining factor*, which stimulates fetal gonadal tissue to develop testes, which secrete *androgen*. The androgen stimulates *embryonic mesonephric ducts* to develop into the male reproductive tract. Lacking the androgens, the embryonic mesonephric duct develops into the *paramesonephric duct*, which becomes the female reproductive tract, and ovaries develop. External genitalia in both males and females begin development as genital tubercles, urogenital folds, and labioscrotal folds. The genital tubercles differentiate into the glans penis in males and clitoris in females. The urogenital folds enclose the urethra in males and develop into the labia minora in females. The labioscrotal folds develop into the male scrotum and into the labia majora in females. The fetal testes descend along the gubernaculum through the inguinal canal into the scrotum shortly before birth or shortly after birth; if it is delayed, the male will likely be infertile. In the female the ovaries descend into the lesser pelvis at the level of the anterior inferior iliac spine prior to birth.

Male

Anatomy

The male anatomy consists of the external *scrotum*, containing the testes and spermatic cord; *ductus deferens;* and the penis. Internally, the spermatic cord passes from the posterior scrotum through the inguinal ring into the pelvic cavity, passing the seminal vesicles, prostate, and bulbourethral glands, into a short ejaculatory duct into the urethra in the penis. The penis has an internal root and the external glans and shaft. Loose skin extends over the sensitive glans, called the prepuce or foreskin. Internally the penis consists of three elongated erectile tissues: the paired dorsal *corpus cavernosum* that, when engorged with blood, create the erection; and a ventral *corpus spongiosum*, containing the urethra. In all three erectile tissues, sinuses called lacunae are separated by connective tissue and smooth muscle (*trabecular muscle*), the lacunae fill with blood upon erection. The proximal end of the corpus spongiosum dilates into a bulb surrounding the urethra and the bulbourethral gland ducts. The corpus cavernosum diverge into paired *crura*, anchoring the penis to the pubic arch and perineal membrane.

Physiology

The production *of gonadotropin* stimulates the beginning of puberty in the male, and its production continues throughout adulthood. The hypothalamus initiates the secretion of *gonadotropic releasing hormone* (GnRH), stimulating the anterior pituitary to secrete *follicle stimulating hormone* (FSH) and *luteinizing hormone* (LH). These hormones stimulate the enlargement of the testes. Luteinizing hormone stimulates the interstitial cells to produce *androgen* (testosterone), influencing the development of secondary sexual characteristics and stimulating indirectly the *spermatogenesis*. Follicle-stimulating hormone stimulates the *sustentacular cells* to produce *androgen-binding protein* (ABP), which binds with the androgen, stimulating sperm production. Secondary sex characteristics are stimulated by GnRH, LH, and FSH along with the androgen. These characteristics consist of body growth with an increase in body muscle mass, increased erythropoiesis, voice changes, and changing hair growth patterns. The libido is stimulated. *Inhibin* secreted by the sustentacular cells selectively inhibits the FSH as a feedback mechanism, controlling sperm production without inhibiting the production of testosterone secretions or LH. At about twenty years of age, the production of testosterone is reduced and stabilizes, along with a decline of the secretory activity of the sustentacular and interstitial cells. The secretion of FSH and LH by the anterior pituitary becomes elevated and may produce physiological changes and mood swings (*andropause*). Erectile dysfunctions may occur at any age without affecting the ability to ejaculate.

Sperm production is temperature sensitive and unable to develop at 37 degrees Centigrade. The testes require a temperature of about 35 degrees or slightly less for successful development of healthy sperm. The temperature reduction is accomplished by the actions of three structures. The *cremaster muscle* in the spermatic cord relaxes to cool the testes, or contracts to warm the testes, maintaining the temperature in an optimal range. The *dartos muscle* in the wall of the scrotum contracts or relaxes, tightening the wall if cool or relaxing the wall to cool the testes. The *pampiniform plexus* of blood vessels act as a countercurrent heat exchanger, cooling the blood entering the scrotum and reducing the temperature to protect the testes. The testis is composed of a fibrous capsule (tunica albuginea) with fibrous septa that divides the testis into 250 to 300 lobules (compartments) that contain one to three sperm-producing seminiferous tubules, while the interstitial cells between the tubules produce the testosterone. The epithelium of the seminiferous tubules contain the germ cells that produce the sperm and the sustentacular cells that produce inhibin, nourish and support the germ cells, and form the blood-testes barrier. The sperm is produced from the germ cells by spermatogenesis, which is the meiotic (reductional) division of the chromosomal number. The twenty-three pairs of chromosomes go through a two-stage process from the primordial germ cells, with twenty-three pairs of genes dividing to yield gametes with twenty-three chromosomes and little else. The primordial germ cells of the seminiferous tubules arise from the embryonic yolk sac and migrate into the gonadal tissue, becoming spermatogonia at birth. When stimulated by hormones at puberty, the spermatogonia begin the divisional process that continues throughout the life of a male.

The divisional process of the spermatogonia results in primary spermatocytes, which enter into meiosis, producing secondary spermatocytes that divide further into spermatids. The spermatids shed excess cytoplasm and produce a flagellum, becoming sperm. The sperm has a head containing the genetic material, an *acrosome* filled with enzymes to aid in penetration of the ova, and a tail composed of mitochondria, nutrients, and contractile elements attached to the flagellum. The ejaculated semen contains about ten percent sperm, thirty percent prostatic fluid, and sixty percent seminal fluid, which contains fructose, *semenogelin* (promotes clotting of the semen), prostaglandins, and *serine protease*, which dissolves the clotted semen and activates the sperm.

Female

The embryonic female is homozygous (XX) for the sex genes. In the absence of the androgen (testosterone) and Müllerian inhibiting factor, the paramesonephric duct develops into the clitoris from the genital tubercle, the labia minora from the urogenital folds, and the labia majora from the labioscrotal folds.

Anatomy

The ovary, like the testes, is derived from the embryonic yolk sac, which migrates and descends into the pelvic cavity. It consists of a central medulla, a cortex that is the surface parenchyma, and a tunica albuginea (fibrous capsule). The ovary is supported by three ligaments. Blood is supplied by the ovarian artery, drained by the ovarian vein, and innervated by the ovarian nerve. Each ovary develops eggs in a bubble-like space (follicle) located in the cortex. When the egg is selected and matures, the follicle in which it develops ruptures, releasing the egg into the pelvic cavity. The *fallopian* (uterine)tube captures the egg in the pelvic cavity by the action of the ciliated cells, which create a low-pressure current to sweep the egg into the tube, transporting it to the uterus. If the egg is fertilized, this occurs in the third of the tube closest to the ovary. The *uterus* is a pear-shaped, thick-walled muscular chamber located superior to the urinary bladder. The regions of the uterus are the upper *fundus*, the *corpus* (body), and *cervix* at the exit into the vagina. The uterine wall consists of an outer layer composed of serosa, called the *perimetrium*; the *myometrium*, which is composed of thick layers of smooth muscle; and an *endometrium*, which is highly vascular and mucosa like. In the endometrium are located numerous tubular glands. It is divided into the *stratum functionalis*, a thick, highly vascular region that is shed during menstruation; and the *stratum basalis*, which regenerates the stratum functionalis after menstruation. The uterus is anchored by four paired ligaments, and blood is supplied by the uterine artery, which arises from the internal iliac artery; ancillary vessels develop during pregnancy. The vagina is attached inferior to the uterus and tilts posteriorly between the urethra and rectum. It lacks glands but it remains moist by serous fluids passing through the vaginal wall by transudation and mucus from the cervical canal.

The external genitalia is collectively called the *vulva* (pudendum) and includes the mons pubis, labia majora, labia minora, clitoris, vaginal orifice, accessory glands (greater and lesser vestibular glands and paraurethral glands), and erectile tissues (vestibular bulbs). The urethra opens in the vulva. The breasts are internally divided into a series of lobes consisting of lactiferous ducts connected to the exterior nipple. Unless pregnant or lactating, the breasts (mammary glands) remain small and nonfunctional. Breast cancer can occur in males and females, and a gene has been implicated as being the potential trigger, although it has been considered nonhereditary. The risk factors for breast cancer are not fully identified.

Physiology

In Western cultures, females begin puberty between nine and ten years of age, when levels of GnRH rise, stimulating the secretion of FSH and LH by the anterior pituitary gland. The FSH stimulates the ovaries to begin secretion of estrogen, progesterone, inhibin, and other androgens. The earliest sign that a young female has begun puberty is beginning of breast development called *thelarche*. This development is stimulated by estrogen, progesterone, prolactin, glucocorticoids, and growth hormone. Pubic and axillary hair development begins at the same time, with the associated sebaceous and axillary sweat gland (*pubarche*), which is stimulated by the androgens, which stimulate libido. Around twelve years of age, the young female experiences the first menstrual period (*menarche*). During the first menstrual cycles, ovulation does not normally occur, so they are said to be anovulatory. Ovulation becomes more regular and predictable after the first year. Estradiol stimulates continued ovarian development, feminizes the external anatomy, and stimulates bone growth. Progesterone stimulates the uterus to become vascular and receptive, and changes in the levels stimulate the menstrual flow. As a female ages, the number of follicles decline, which reduces the production of estrogen and progesterone. Aging brings a time of transition (climacteric) that lasts for several years before menopause occurs with the cessation of ovulation and menstruation.

Oogenesis

The production of ova begins the fifth month of fetal development as the oogonia multiply. Some oogonia develop into primary oocytes, which undergo the first stage of meiosis prior to birth. Most of these primary oocytes are lost by *atresia* in the fetus and before puberty, with about 400,000 remaining as puberty begins. The surviving oocytes complete meiosis I, dividing into a large secondary oocyte and a small, nonviable polar body. The secondary oocyte enters meiosis II, proceeding to the metaphase II stage where further development ceases unless fertilization occurs. Fertilization triggers the continuation of meiosis II, with unequal division releasing a second polar body and a zygote, which will become the fetus. This development occurs in a follicle in the ovarian cortex. The follicle consists of the primary oocyte and a single layer of squamous follicular cells. With development the follicular cells become cuboid and develop into the secondary follicle. These follicular cells become stratified, forming the tertiary follicle with pools of follicular fluid that merge to form the antrum (large central cavity). As the follicle matures, multiple layers of follicular cells (*granulosa cells*) cover the ova (called *cumulus oophorous*), securing the ova to the wall of the follicle. Each month, one or more tertiary follicles mature and rupture in ovulation. The inner layer of the cumulus cells form the *corona radiat*, surrounding the ova and separated from the ova by the gelatinous glycoprotein zona pellucida.

The ovarian cycle occurs in the absence of pregnancy, lasting approximately twenty-eight days from the first day of menstrual flow. Following ovulation, a new cohort of follicles begins to develop, with one being the dominant follicle, which ruptures two months later. The follicular phase is stimulated by FSH as well as the follicular development. The follicles that begin development and are not dominant undergo atresia (reabsorption). Ovulation occurs around the fourteenth day of the cycle, stimulated by decrease of LH in varying concentrations. The follicle that has ruptured becomes the *corpus hemorrhagicum*, which rapidly converts into the *corpus luteum* (yellow body), producing an increased concentration of *progesterone*, creating the luteal phase that ends at menstruation. Until the twenty-second day of the cycle the increased progesterone and estradiol continue, and begin to taper off unless fertilization occurs and implantation has been initiated. The corpus luteum becomes the *corpus albicans* at about the twenty-sixth day, ceasing all progesterone production, which leads to the breakdown of the stratum functionalis of the uterus over the next two days, beginning the next cycle.

The menstrual cycle is divided into four phases. The first is the *proliferation phase*, which occurs between the sixth and fourteenth day of the cycle as the uterus stratum functionalis is repaired by the stimulation of estrogen. The second is the *secretory phase*, between the fifteenth and twenty-sixth day of the cycle, with the thickened endometrium secreting mucus and glycogen to prepare for implantation; this is stimulated by progesterone produced by the corpus luteum. The *premenstrual phase* occurs between the twenty-sixth and twenty-eighth days, with the reduction of progesterone and the conversion of the corpus luteum into the nonfunctional corpus albicans, resulting in the cessation of extra blood flow to the endometrium, resulting in necrosis of the stratum functionalis. The *menstrual phase* between the twenty-eighth and fifth day of the cycle begins with the sloughing of the necrotic stratum functionalis, with associated capillary bleeding producing the vaginal discharge during menstruation.

Pregnancy

The gravid state (pregnancy) lasts approximately 266 days from the date of conception (fertilization of the ova) to birth (parturition), or 280 days from the beginning of the last menstrual period. The embryo (fetus) and associated membranes are the conceptus. Within twenty-four hours of the beginning of implantation, which begins as the blastula of cells created from the fertilized ova burrowing into the stratum functionalis, the production of *human chorionic gonadotropin* (HCG) begins by the uterus. The HCG stimulates the corpus luteum to continue its function of producing estrogen and progesterone. The elevated levels of progesterone and estrogen support the development of the placenta and fetus for the first trimester. By the end of the first trimester the placenta is adequately developed, assuming the role of the corpus luteum and increasing the production of estrogen and progesterone. The estrogen stimulates tissue growth of the mother and the softening of the sacroiliac joint and pubic symphysis in the pelvic girdle to ease vaginal birth. The progesterone inhibits uterine contractions and promotes mitosis of uterine cells to nourish the fetus. Both estrogen and progesterone stimulate mammary development and inhibit FSH secretion. Human chorionic somatomammotropin (HCS) produced by the placenta mobilizes fatty acids for fuel for the mother, sparing glucose for the fetus.

During pregnancy, the mother experiences many discomforts: morning sickness occurring early while the hormone levels are escalating; constipation due to physical and fluid changes; and many others. The mother's basal metabolic rate increases, with associated hot flashes, and increased nutrient intake to meet the fetal demands. Blood volume and cardiac output increase to meet fetal demands and growth. Ventilation changes over the pregnancy to increase the available oxygen and eliminate carbon dioxide. Urinary output increases to eliminate the additional wastes produced by the fetus and mother. The mother's skin stretches and frequently tears with the scarring (striae) of the abdomen and breasts. Darkened areas develop with increased melanin production on the face and abdomen. During the seventh month of gestation, the fetus may turn head down (vertex) and the mother will experience unorganized contractions of the uterus (*Braxton Hicks*) before true labor. As term approaches, the posterior pituitary increases the secretion of *oxytocin*. The production of estrogen and progesterone changes, with the uterus becoming more contractile. The cervix is stretched by the uterine contractions, thinning (*effacement*), and the *cervical os* dilates to ten centimeters, which is the first stage of labor, called dilation. The fetal membranes rupture, releasing amniotic fluid (water breaking). The baby's head enters the vagina, indicating the second stage (expulsion), which lasts until the baby is fully discharged, with the clamping and cutting of the umbilical cord. The *placental delivery* is the third stage, with the expulsion of the placenta, amnion, and other components, clearing the uterus. The next six weeks postpartum is called *puerperium.* This is marked by discharge of fluids and cellular debris called *lochia*, contraction of the uterus, and other readjustments to a pregravid state.

Lactation

During pregnancy the mammary glands complete development, stimulated by increases in estrogen, progesterone, and prolactin in preparation for lactation. This causes the secretory acini connected to the ducts to develop fully and increase in number. Once birth occurs, for the next three days the mammary glands secrete colostrum, which is a rich source of antibodies as well as nutrients for the infant. The *colostrum* is richer in proteins than milk, lower in fats and lactose, and contains immunoglobulin to provide immunity to infection. After about three days milk production begins to support the infant. The delivery of the placenta stimulates the secretion of additional prolactin, supporting lactation.

CHAPTER THIRTEEN
THE IMMUNE AND LYMPHATIC SYSTEM
ORGANIZATION AND FUNCTION OF FLUID RECOVERY, PREVENTION, AND DEFENSE AGAINST DISEASES

LEARNING OBJECTIVES

Upon completion of the chapter readers will have a basic understanding of the lymphatic and immune systems:

1. Anatomy and organization of the lymphatic system and its interaction with the cardiovascular system
2. Role of the lymphatic and immune components in the monitoring and protection of the body
3. Organization of cellular immunity and the use and production of antibodies
4. Immunoglobulins and immune responses

The immune and lymphatic system are closely interrelated in description and function. The functional lymphatic system provides the stimulus and antibodies to the immune system, triggering its response. Many of the immune responses occur in the lymphatic structures, or the lymphatic structures remove the products of the immune system.

Lymphatic System

The lymphatic system functions to recover and filter tissue fluids to the cardiovascular system; provide the cellular components to the immune system; monitor the body fluids for the presence of foreign cells, toxins, and atypical body cells; and transport dietary lipids from the small intestine to the liver and blood. The system is composed of the lymph fluid, lymphatic capillaries and veins, lymphatic tissues (nodes) and the lymphoid organs (tonsils, spleen, and thymus). The lymph is the colorless interstitial fluid produced from the blood plasma and not recovered by the blood capillaries. The lymph from the small intestines usually is milky-looking due to the large amount of lipids. Lymph contains lymphocytes, wandering macrophages, hormones, cellular debris, bacteria, viral particles, and metastasizing cells. Lymph originates when the blind capillaries of the lymphatic system collect the tissue fluids and its contents from the surrounding tissues through large pores in the capillary wall, aided by muscle contraction. Lymphatic capillaries are larger than those in the cardiovascular system, and blind, meaning there is only a connection on one end to the lymphatic vessels (veins). The lymphatic capillaries form regional drainage systems to recover the tissue fluids. They merge into collecting vessels connected to lymph nodes, where filtration of the fluid occurs, and converge to form regional lymphatic trunks that merge into the thoracic and right lymphatic collecting ducts, which empty into the subclavian veins, returning the fluid to the cardiovascular system. Lymphatic vessels are like veins in that they have directional flow provided by muscular activity and valves in the vessels.

Lymphatic Cells

Lymphatic cells are a variety of lymphocytes and associated cells. *Natural killer* (NK) cells provide immune surveillance; *T lymphocytes* are lymphocytes maturing in the thymus; *B lymphocytes*, which provide memory and convert into plasma cells, producing antibodies; *macrophages*, which phagocytize various debris and present antigens to stimulate immune response; *dendritic cells* found in the epithelial membranes and lymph nodes, acting as antigen-presenting cells; and *reticular cells,* found in the stroma of the lymphoid organs and acting as antigen-presenting cells in the thymus. Diffuse lymphatic cells are found in mucous membranes of the respiratory, digestive, urinary, and reproductive systems, which form *mucosa associated lymphatic tissues* (MALT). Dense concentrations of lymphatic cells and nodules are found in the ileum of the small intestine (Peyer patches), controlling antigen entry into the body from the alimentary system.

Lymphatic Organs

Primary lymphatic organs are the *red bone marrow* and the *thymus*, where lymphocytes mature before colonizing other sites, which are *secondary lymphatic organs* like the *spleen, tonsils*, and *lymph nodes*. The red bone marrow consists of hemopoietic islands composed of macrophage and developing blood cells separated from other islands by sinusoids, which converge into a central vein. This central vein permits the mature cells to enter the bloodstream. The thymus is located in the mediastinum, along with the heart. The T lymphocytes develop from undifferentiated lymphocyte stem cells from the red bone marrow, which are transported to the thymus and are isolated during differentiation and development. The thymus is composed of dense cortex divided into lobules and a lighter medulla. The cortex is separated from the medulla by reticular epithelial cells surrounding the blood vessels forming the *blood-thymus barrier* isolating the developing T- lymphocytes from blood-borne antigens. Lymph nodes are numerous small, bean-shaped structures connected to the lymphatic vessels and capillaries from afferent vessels. The nodes filter the lymph and pass the lymph further through efferent vessels. The lymph in the node is monitored for foreign antigens or other atypical materials. The functional part of the lymph node has an outer cortex with nodules containing the T lymphocytes and a medulla where the B lymphocytes are located. Lymph nodes are concentrated in the cervical, axillary, and thoracic areas of the upper body, and in the abdominal (intestinal and mesenteries), inguinal, and popliteal group in the lower body. The tonsils, located in the pharynx, are lymphatic organs that guard against airborne or ingested pathogens. The tonsils are located in three clusters: in the nasopharynx medial *pharyngeal tonsils*; paired *palatine tonsils* in the rear of the oral cavity; and *lingual tonsils* located at the root of the tongue. Tonsils are composed of a superficial epithelium and a deeper fibrous capsule, with deep pits called *tonsillar crypts*, along with rows of numerous lymph nodules. The spleen is the largest lymphatic organ, located between the diaphragm, kidney, and stomach. It is divided into a *red pulp*, containing concentrated erythrocytes, and a *white pulp*, containing lymphocytes and macrophages. The spleen monitors the blood for foreign antigens, and regulates blood volume as a reserve site for platelets and erythrocytes.

Immunity and Immune System

Immunity is divided into non-specific resistance (immunity), involve a variety of nonspecific reactions to reduce the development of infections; and specific immunity, which involves reactions by the immune system in response to identified antigenic threats that stimulate a directed response to defeat the threat.

Nonspecific Resistance/Immunity

Nonspecific immunity consists of three lines of defense to protect the body. The first is the external physical barriers provided by the *skin* and mucous membranes, which block invasion of the body by foreign organisms. *Nonspecific defense* occurs when the pathogens breech the physical barriers and the body responds with actions of macrophages, lymphocytes, antimicrobial proteins, and fever. The third line of defense is the immune system, which uses all capabilities in the second line of defense to defeat the invader and retains a memory for future encounters. The skin is an effective physical barrier, with the low moisture content, keratinized cells, fatty acids, and defensins present on the surface. The mucous membrane is effective with surface containing agents that reduce adherence and constant moisture to flush the surface, which contains hyaluronic acid and lysozymes. Neutrophils, the most abundant leucocytes, phagocytize bacteria and produce stronger oxidizers, which destroy the ingested invaders. Eosinophils produce agents that interfere with parasites and promote basophils and mast cells activity, controlling inflammatory response. Basophils secrete leukotrienes, histamines, and heparin, promoting increased blood flow and attracting leucocytes into the area; this includes NK lymphocytes. Monocytes differentiate into macrophages that phagocytize foreign material and present the antigens, stimulating immune activity. Specific macrophages are associated with various locations, with microglia acting as phagocytes in the cerebrospinal fluid, dust cells in the alveoli, and hepatic macrophages in the liver. The dendritic cells associated with the skin have a limited phagocyte capability.

Interferons, which are polypeptides secreted by cells in response to viral invasion, alert neighboring cells to produce antiviral proteins to block further invasion. NK lymphocytes are activated to move to the invaded cells and begin destruction of the cells.

The *complement system,* which consists of thirty or more globulins, is activated by pathogens being present. This initiates an enhancement of the inflammatory process, opsonizing of the bacteria inactivating them, and stimulating immune clearance of the antigens and the cytolysis of the foreign cells.

Immune surveillance by the NK lymphocytes detects and destroys foreign and diseased host cells by secreting perforins, which make the cell membrane or wall porous for the loss of cytoplasm and the secretion of granzymes, leading to cell death.

Fever (*pyrexia*) is induced by pyrogens introduced by bacteria or virus particles, secreted by neutrophils and macrophages to stimulate a rise in the metabolic activity in the area or body above the normal set point to inhibit pathogenic reproduction or spread. Local increases in temperature are related to the leucocytes, while rise in the body is stimulated by the hypothalamus.

Inflammation is the response to an infection or trauma that is characterized by redness, swelling, increased temperature, or pain. This begins the mobilization of defenses by releasing of histamines, leukotrienes, and other cytokines, triggering local vasodilation and increasing capillary permeability. An increase of chemicals and defensive cells occurs, with leucocytes adhering to vessel walls (*margination*) and then slipping between the endothelial cells of the vessels into the tissue spaces (*diapedesis*). These leucocytes are attracted to the damaged site by chemotaxis. The inflammatory process contains the infective agents or injury site. This allows the focus of the defense and repair to occur. This includes the removal of damaged tissue and clotting; phagocytic activity and antibodies to remove the infective agents; increased fluid in the tissue (*edema*), reducing mobility; and increased lymphatic drainage once the tissue repair is underway.

Specific Immunity

Specific immunity is based on cell memory and the development of specific responses to pathogens. Specific immunity is driven by cellular activity (*cellular immunity*) or by antibodies (*humoral immunity*). Immunity can be based on active development of cells and antibodies by the body upon exposure to antigenic materials; passive, where the antibodies or lymphocytes are provided from another source; or artificial, through administration of immune sera or vaccines. Memory only develops from active contact with antigenic material to provide long-term protection.

Antigens are organic molecules that are capable of being recognized as unique and inducing an immune response. These are usually large polysaccharides, glycoproteins, or glycolipids. The active sites that trigger the antigenic properties are called *epitope*. Some very small molecules called *haptens* can combine with larger host molecules, resulting in antigenic characteristics developing.

T lymphocytes mature in the thymus, surviving by negative selection, and migrate to populate the lymph tissues and organs. B lymphocytes mature in the bone marrow, surviving negative selection to, also, populate lymph tissue and organs. The *antigen-presenting cells* are dendritic or macrophage, reticular cells and B lymphocytes that process the antigens derived from other sources and display epitopes on the cell surface. There is the *major histocompatibility complex* (MHC), which consists of proteins that alert the immune system to the presence of foreign antigens or materials. Chemical signals used for communication between immune cells are the *interleukins*.

Cellular immunity

Cellular immunity employs four classes of T lymphocytes to accomplish defense of the body. They are the *cytotoxic T cells*, which use cytotoxic chemicals to destroy foreign antigenic materials (cells); *helper T cells*, which aid in the recognition of foreign antigens binding to the antigenic source, acting as a maker for the cytotoxic T cells; the *regulatory T cells*, which trigger clonal selection and multiplication of the T cells to attack the invaders; and some regulatory T cells are retained becoming *memory T cells*. Attack of the foreign antigens is triggered by the secretion of interleukins attracting neutrophils, NK cells, and macrophages. In the rapid multiplication and maturation of B lymphocytes and T lymphocytes, cytotoxic T lymphocytes are promoted. This permits increased destruction of the targeted cells having the antigens. The memory cells developed by the first antigenic exposure provide the history of the primary response, which allows a more rapid activation of immune system with another, later exposure.

Humoral immunity

Humoral immunity is based on the production of antibodies rather than cellular recognition, attack, and memory. B lymphocytes that *immunocompetent* recognize and internalize the antigen, displaying the epitope on the surface with MHC- II; helper T lymphocytes bind with the antigen, secreting interleukins and activating B lymphocytes. This stimulates rapid and repeated division of the B lymphocytes, with some becoming *memory B lymphocytes* while others become *plasma cells.* The plasma cells synthesize antibodies that inactivate the antigens *by neutralization, complement fixation, agglutination,* or precipitation. These antibodies are *immunoglobulins*, which have two sites capable of binding with the specific antigen. There are five principle immunoglobulins listed as immunoglobulin A, located in the plasma; immunoglobulin D, located in B lymphocytes; immunoglobulin E, located inbasophils and-mast cells; immunoglobulin G, the most abundant in plasma; and immunoglobulin M, located in B cells.

Immune Disorders

Hypersensitivity is when the body has an excessive reaction when exposed to an antigen, which can be tolerated by most individuals.

- Type I is an acute reaction with an immunoglobulin E mediated response, which begins within seconds of exposure and usually subsides within about thirty minutes when treated (asthma, anaphylaxis, anaphylactic shock).
- Type II, antibody-dependent cytotoxic hypersensitivity, is mediated by immunoglobulins G or M, which promote the attack of the antigen bound on target cells like the erythrocytes, which promote transfusion reactions.
- Type III, an immune complex hypersensitivity reaction, is a widespread antigen antibody complex formation that triggers intense inflammatory response, as encountered in glomerulonephritis and systemic lupus erythematosus.
- Type IV is a delayed hypersensitivity reaction, appearing twelve to seventy-two hours after exposure to the antigen, mediated by cellular reaction to the antigen such as seen in poison ivy dermatitis and TB skin tests.

Autoimmune reactions are failures of the immune system to identify antigens on the body's cells and distinguish them from foreign antigens, which results in the immune system attacking the body and destroying itself. This is normally a failure of the regulatory function of T lymphocytes and may have cross reactions from rheumatic fever or type I diabetes.

Immunodeficiency diseases are failure of the immune system to respond to antigens, so that it is incapable of defending the body from foreign invaders. Severe combined immunodeficiency disease (SCID) can affect newborns. Individuals with AIDS, are infected by a retrovirus that destroys thymus capability to produce the helper T lymphocytes, permitting opportunistic infections which can prove fatal.

CHAPTER FOURTEEN

THE ENDOCRINE SYSTEM

ORGANIZATION AND FUNCTION OF HORMONES

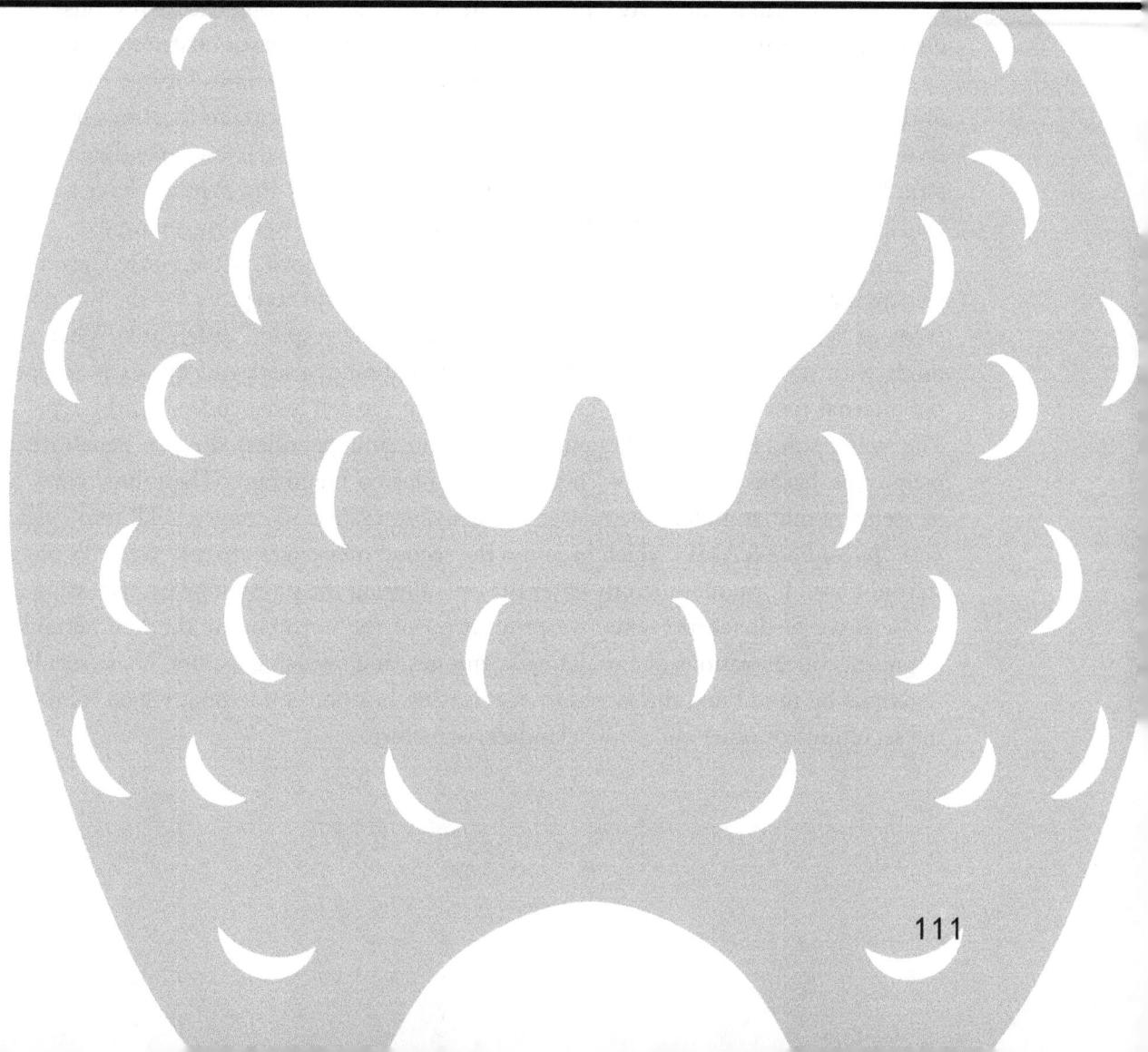

LEARNING OBJECTIVES

Upon completion of the chapter readers will have essential and basic understanding of the endocrine system and the role of hormones in the body:

1. The location, organization, and functions of the endocrine glands
2. The impact of hormones on receptive tissues and their interactions within the body

The endocrine system, working in concert with the nervous system, maintains the internal environment of the body. While the nervous system's control is rapid and short acting, the endocrine system's is slower, through chemical messengers called hormones, and longer lasting. These hormones stimulate changes in metabolic activities and functions throughout the body in any tissue that has receptors for the specific hormone. Hormonal changes control the physiology in five broad actions: they regulate the composition and volume of the internal environment; regulate the metabolic rate and energy production; adjust the environment in response to emergency demands; coordinate growth and development in the body; and stimulate reproduction and repair.

The endocrine system consists of a series of glandular structures located in various locations in the body that produce the hormones that the blood transports to locations in the body where receptors are located. These hormones are grouped into three classes, depending on the molecular structure. Amines are developed from amino acids; proteins and peptides have a series of amino acids; and steroids are developed from cholesterol or other sterol-ringed lipids.

Hormones are released on demand in amounts determined by the body. They are targeted to principal action sites, having target cells with appropriate receptors for the hormone. Different types of cells having receptors for the same hormone respond differently. When the hormone binds with its receptor on the cell's plasma membrane, a series of events is initiated, altering the internal functions of the cell. The response of the cell is dependent on the type of hormone. The water-soluble hormones (amines, proteins, and peptides) do not penetrate the plasma membrane because of the lipid layers in the plasma membrane. These hormones are the first messengers that stimulate the release of adenylate cyclase to convert ATP into *cyclic adenosine monophosphate* (cAMP) which becomes the second messenger altering the cell's physiology. The steroid-based hormones directly enter the cell, altering the physiology by activating genes.

Negative feedback prevents overproduction of the hormone by the interactions of nervous impulses, concentrations of circulating hormones, and releasing factors. These regulate the hypothalamus by modifying the secretion of releasing hormones (factors), which inhibit or promote the secretions of other endocrine glandular structures.

Glands

The primary gland of the endocrine system is the *pituitary gland*, it interacts with the hypothalamus to secrete hormones that initiate and stimulate the activities of other glands. The pituitary gland is located in the *sella turcica* in the cranium and is divided into a posterior segment and anterior segment connected by the pars intermedia.

The *anterior pituitary gland* is considered to be the master gland. It releases a variety of hormones that regulate functions. The secretions of the anterior pituitary gland are regulated, inhibited, or stimulated by regulating hormones from the hypothalamus. Seven hormones are secreted by the anterior pituitary gland. These hormones are *human growth hormone* (HGH), which promotes cellular reproduction and repair; *prolactin* (PRL), which activates the development of mammary tissues; *adrenocorticotropic hormone* (ACTH), which stimulates the adrenal cortex to produce androgens, mineralocorticoids, and glucocorticoids; *melanocyte stimulating hormone* (MSH), which promotes the darkening of the skin; *thyroid stimulating hormone* (TSH), which stimulates the actions of the thyroid gland; *follicle stimulating hormone* (FSH), which stimulates gonad activities; and *luteinizing hormone* (LH), which stimulates gonad activities.

The *posterior pituitary gland* is primarily a storage site for hormones produced by the hypothalamus. *Oxytocin* (OT) is a hormone that stimulates smooth muscle contraction intensity. *Antidiuretic hormone* (ADH) promotes increased permeability of the renal tubules.

The *thyroid gland* is located inferior on either side of the larynx. It is divided into two lobes consisting of follicular cells that secrete the thyroid hormones *thyroxine* (T4) and *triiodothyronine* (T3), which regulate the metabolism of organic molecules to control energy balance, growth, and development, and activate the nervous system. Parafollicular cells secrete *calcitonin* (CT), which controls the calcium levels in the blood by stimulation of deposition of calcium and phosphates in bone tissue.

The *parathyroid gland* is located in paired follicles on the posterior surface of the thyroid gland, consisting of chief cells and oxyphil cells that secrete *parathyroid hormone* (PTH). Parathyroid hormone regulates calcium and phosphate levels in the blood by stimulating the breakdown of bone, increasing the calcium levels; promoting the recovery of calcium from the filtrate in the nephrons; and blocking the recovery of phosphates from the filtrate, functionally increasing the calcium level in the plasma and reducing the phosphate levels.

The *adrenal glands* are located superior to the kidneys. They consist of an outer medulla that synthesizes and secretes *epinephrine* and *norepinephrine* in support of the autonomic nervous system's sympathetic division; and an inner cortex, divided into three zones. The outer zone of the cortex secretes mineralocorticoid hormones; the middle zone secretes hormones called glucocorticoids; and the inner zone secretes androgens. The mineralocorticoids, principally *aldosterone*, promote the reabsorption of sodium and water from the renal tubules, regulating the sodium-potassium balance in the blood. The secretion is controlled by the renin-angiotensin pathway, which is triggered by a decrease in the mean arterial blood pressure and decreases of sodium ions and extracellular fluids, which promote renin secretion by the kidney. The glucocorticoids consist of *cortisone* and *cortisol*, which promote organic molecular metabolism in response to stress and serve as anti-inflammatory agents. The androgens are a combination of *estrogen* and *testosterone,* which are at low levels before puberty and enable development of secondary sex characteristics during puberty.

The *pancreas* is a dual-function gland, providing enzymes to the digestive system and hormones to control physiological management of blood glucose. *Insulin* is produced in the beta islets of Langerhans. It reduces blood glucose levels by promoting absorption of glucose by the cells, converting glucose to glycogen in the muscles and liver, and converting excess glucose into lipids. *Glucagon* is produced by the alpha islets of Langerhans. It raises the blood glucose by stimulating the breakdown of glycogen in the liver, producing glucose during periods over six hours without meals, to maintain a level blood glucose level. *Growth-inhibiting hormone* is produced by the delta islets of Langerhans. It acts to inhibit the production and secretion of insulin and glucagon.

The gonads, *ovaries* and *testes,* produce the sex hormones, beginning shortly after conception and becoming inactive until re-stimulated at puberty. The ovaries are located in the pelvic cavity, producing the primary hormones *estrogen* and *progesterone,* and secondary hormones *inhibin* and *relaxin.* These hormones control the primary and secondary sex characteristics of the female, the menstrual cycle, pregnancy, lactation, and other reproductive functions. The testes of the male are located in the scrotum and produce the primary hormone *testosterone* and secondary hormone *inhibin.* The primary hormone is responsible for the development of the sexual characteristics of the male and reproductive functions.

The *pineal gland* is located attached to the roof of the third ventricle near the hypothalamus. It accumulates calcium at puberty and secretes *melatonin,* which inhibits reproductive activities by the inhibition of release of gonadotropic hormones.

The *thymus* is a glandular site located near the heart in the mediastinal space. It produces a series of hormones involved with the production, maturation, and activation of T lymphocytes.

There are numerous other tissues that have endocrine-like functions, secreting hormones to control local activities or to stimulate continuing function of other glandular structures. In the gastrointestinal tract, tissues secrete *gastrin, secretin, cholecystokinin, enterocrinin, villikin,* and *gastric inhibitory peptides,* which are distributed through the blood and affect local actions; some affect sites removed from the gastrointestinal tract. The secretion of chorionic gonadotropin by the lining of the uterus upon the beginning of implantation stimulates the continued function of the corpus luteum through the first trimester. Inadequate oxygen detected by the efferent arterioles of the nephron stimulates the kidney to secrete erythropoietic factor, promoting increase in the production of erythrocytes in the red bone marrow. The stretching of the atria by increased blood volume stimulates the secretion of *atrial natriuretic peptide,* which inhibits aldosterone, decreasing sodium levels in the blood and increasing the urinary output, reducing fluid volume.

APPENDIX

TABLES WITH SUMMARY DATA OF INTEREST

Table 1. A summary of the cellular components, description, and function

COMPONENT	STRUCTURE	FUNCTION
Plasma membrane	Lipid bilayer with scattered proteins and cholesterol molecules	Maintains the cell's boundary and integrity Embedded proteins have multiple functions
Nucleus	Surrounded by double-layered nuclear envelope (membrane)	Houses chromosomes that possess DNA, which dictates cellular function along with protein synthesis
Nucleolus	Located inside the nucleus, appears dark with an oval shape	Location of the synthesis of ribosomal RNA
Cytosol	Gel-like fluid within the cell	Location of cell metabolism and storage
MEMBRANE-BOUND ORGANELLES		
Rough endoplasmic reticulum	Continuous with the nuclear envelope; flattened sacs dotted with ribosomes	Location of protein synthesis and post-translational processing
Smooth endoplasmic reticulum	Continuous with the rough endoplasmic reticulum; tubular structure without ribosomes	Lipid synthesis and post-translational processing; transport of molecules from the endoplasmic reticulum to Golgi apparatus; calcium storage
Golgi apparatus	Series of flattened sacs near the endoplasmic reticulum	Post-translational processing; packaging and sorting of proteins
Mitochondria	Oval shaped with an outer membrane and inner membrane, with folds called cristae projecting into the matrix	ATP synthesis
Lysosomes	Granular, sac-like; scattered throughout the cytoplasm	Breakdown of cellular and extracellular debris
Peroxisomes	Appear similar to lysosomes but smaller	Breakdown toxins, including peroxides
NONMEMBRANE ORGANELLES		
Vaults	Small, barrel-shaped	Unknown; possibly transport between nucleus and cytoplasm
Ribosomes	Granular shaped, composed of proteins and rRNA located in cytosol or on the surface of rough endoplasmic reticulum	Translation of mRNA to synthesize proteins
Centrioles	Two cylindrical bundles of protein filaments perpendicular to each other	Directs mitotic spindle development during cell division
Cytoskeleton	Composed of protein filaments, includes microfilaments, intermediate filaments, and microtubules	Supports the structure of the cell, cell movement and contraction

Table 2. Summary description of epithelial tissues

EPITHELIAL	DESCRIPTION	LOCATION (EXAMPLE)	FUNCTION
Simple squamous	Single layer of squamous cells; shaped like fried egg	Alveoli of lungs; glomerular capsule in kidney; endothelium of blood vessels; serous membranes	Allow rapid diffusion; secrete serous fluid
Simple cuboidal	Single layer of square or round cells	Liver; thyroid; mammary; salivary glands; tubules of kidney and bronchioles	Absorption and secretion; produce mucus; movement of respiratory mucus
Simple columnar	Single layer of tall, narrow cells; nucleus located in the basal half of the cell; may be ciliated, have microvilli, or be associated with goblet cells	Lining of stomach, intestines, gallbladder, uterus, and uterine tubes; some kidney tubules	Absorption, secretion of mucus; movement of ova and embryo in uterine tubes (fallopian, oviduct)
Ciliated pseudostratified columnar	Appear multilayered; composed of mature and immature cells near basement membrane; frequently associated with goblet cells and ciliated	Upper and lower airway from nasal cavity to the bronchi; portions of male urethra	Secretion of mucus for entrapment of foreign particles and to propel mucus from the airway
Stratified squamous keratinized	Multiple cell layer with the cells becoming more flattened at the surface, which is composed of compact dead cells; may appear cuboidal in the basal area	Epidermis, palms of hands and soles of feet are heavily keratinized	Resists abrasion and penetration by pathogens; retards water loss or gain through the skin
Non-keratinized	As above, lacking the dead compact surface cells	Tongue; oral mucosa; esophagus; anal canal; vagina	Resists abrasion and penetration by pathogens
Stratified cuboidal	Two or more layers of round or square cells	Sweat glands; ova-producing vesicles, sperm-producing ducts (seminiferous tubules) of testis	Secretion of sweat in thermoregulation; secretion of ovarian hormones and sperm production
Transitional	Resemble stratified squamous but have more rounded surface; appear to flatten, becoming thinner as stretched, with some cells having two nuclei	Urinary tract; part of kidney, ureter, bladder, urethra, and umbilical cord	Stretches to allow filling, particularly the urinary bladder

Table 3. Summary of the various connective tissues

CONNECTIVE	DESCRIPTION	LOCATION (EXAMPLE)	FUNCTION
Fibrous areolar	Loose arrangement of collagen and elastic fibers; scattered cells; abundant ground substance (matrix); numerous blood vessels	Underlies epithelia; surrounds blood vessels, nerves, esophagus, and trachea; fascia between muscles, mesenteries; visceral layer of serous membranes	Loosely bind epithelia to deeper tissues; passage of nerves and blood vessels through other tissue; provide area for immune defense; provide nutrients from blood and waste removal from epithelia
Reticular	Loose network of reticular fibers and cells; infiltrated with leucocytes (lymphocytes)	Lymph nodes, spleen, thymus, bone marrow	Framework of lymphatic organs
Dense regular	Densely packed collagen fibers (maybe wavy) with slender fibroblasts; nuclei compressed between collagen bundles with little open space	Tendon and ligaments	Tightly bind bones together; resist stretching (stress) for ligaments; tendons attach muscle to bone, transferring tension to the bones to enable movement
Dense irregular	Densely packed collagen fibers in random directions; very little open space and few visible cells	Deeper dermis; capsules around viscera (liver, kidney, spleen) and fibrous sheaths around cartilage and bones	Durable, withstand stresses applied in unpredictable directions
Hyaline cartilage	Clear, glassy matrix; fine dispersed collagen fibers; cells (chondrocytes) enclosed in lacunae in clusters; usually covered by perichondrium	Thin articular cartilage that lacks perichondrium over the ends of movable joints; supports rings of trachea and bronchi; fetal	Eases joint movement; keeps airway open during ventilation (respiration); moves vocal cords during speech; bone precursor in fetus; growth zones in long bones
Elastic cartilage	Elastic fibers form web-like mesh amid lacunae; always covered by perichondrium	External ear; epiglottis	Provides flexible and elastic support
Fibrous cartilage	Parallel collagen fibers like tendon; rows of chondrocytes in lacunae between collagen; never has perichondrium	Pubic symphysis; intervertebral discs; menisci; at tendon insertion on bones	Resists compression, absorbing shock

| Osseous (bone) | Mineral (calcified) matrix arranged in concentric patterns (lamellae) around a central canal in compact bone; osteocytes housed in lacunae between adjacent lamellae; lacunae connected to central canal and each other by delicate canaliculi (little canals) | Compact bone of skeleton | Physical support of body; leverage for muscle action; protective enclosures of viscera and reservoir of calcium and phosphorous |
| Vascular (blood) | Appear as diversified cells (formed elements) loosely associated with each other; most numerous are red blood cells lacking a nucleus interspersed with various white blood cells; sometimes with cell fragments (platelets) | Contained in the heart and blood vessels; spleen, liver | Transport gases, nutrients, wastes, chemical signals and heat; provide defense (leucocytes); contain clotting agents; promote tissue maintenance and repair |

Table 4. Normal value ranges in blood plasma

COMPONENT MEASURED	NORMAL VALUES	NOTES
PHYSICAL FACTORS		
Blood volume	80 to 85 ml/kg of body weight	
Blood osmolarity	285 to 295 mOsm	
Blood pH	7.38 to 7.44	
ENZYMES		
Creatine phosphokinase (CPK)	Females: 10 to 79 U/L	
	Males: 17 to 148 U/L	
Lactic dehydrogenase (LDH)	45 to 90 U/L	
Phosphatase (acid)	Females: 0.01 to 0.56 Sigma U/ml	
	Males: 0.13 to 0.63 Sigma U/ml	
HEMATOLOGY		
Hematocrit	Females: 36% to 46%	
	Males: 41% to 53%	
Hemoglobin	Females: 12 to 16 g/100 ml	
	Males: 13.5 to 17.5 g/100 ml	

(Continued)

Table 4—Continued. Normal value ranges in blood plasma

COMPONENT MEASURED	NORMAL VALUES	NOTES
Red blood cell count	4.50 to 5.90 million/mm^3	
White blood cell count	4.5 to 11.0 thousand/mm^2	
HORMONES		
Testosterone	Males: 270 to 1,070 ng/100 ml	
	Females: 6 to 86 ng/100 ml	
Adrenocorticotrophic hormone (ACTH)	6 to 76 pg/ml	
Growth hormone (GH)	Children: over 10 ng/ml	
	Adult male: below 5 ng/ml	
Insulin	2 to 20 microU/ml (fasting)	
IONS		
Bicarbonate	24 to 30 mmol/l	
Calcium	9.0 to 10.5 mg/dl	
Chloride	98 to 106 mEq/L	
Potassium	3.5 to 5.0 mEq/L	
Sodium	135 to 145 mEq/L	
ORGANICS (NUTRIENTS AND WASTES)		
Cholesterol	Under 200 mg/dl	
Glucose	75 to 115 mg/dl (fasting)	
Lactic acid	5 to 15 mg/dl	
Protein (total)	5.5 to 8.0 g/dl	
Triglyceride	Under 160 mg/dl	
Urea nitrogen (BUN)	10 to 20 mg/dl	
Uric acid	Males: 2.5 to 8.0 mg/dl	
	Females 1.5 to 6.0 mg/dl	

Table 5. Summary of the formed elements of the blood

COMPONENT	DESCRIPTION OF COMPONENT	NUMBERS	FUNCTIONS
Erythrocytes (red blood cells; RBCs)	Biconcave, without nucleus or mitochondria; contains hemoglobin; survives 100 to 120 days	4 to 6 × 10^6	Transport of oxygen and carbon dioxide
Leukocytes (white blood cells; WBCs)			
Granulocytes	Approximately twice the size of RBCs; cytoplasmic granules; survive 12 hours to 3 days		
Neutrophils	Nucleus with 2 to 5 lobes; cytoplasmic granules stained slightly pink	54% to 62% of WBCs	Phagocytic (microphages)
Eosinophils	Nucleus bi-lobed; cytoplasmic granules stain red with eosin	1% to 3% of WBCs	Detoxify foreign substances; secrete enzymes that dissolve clots; fight parasitic infestations
Basophils	Nucleus lobed; cytoplasmic granules stain blue with hematoxylin	Less than 1% of WBCs	Releases anticoagulant heparin
Agranulocytes	Cytoplasmic granules not apparent; survive 100 to 300 days (some much longer)		
Monocytes	2 to 3 times larger than RBCs; nucleus varies in shape	3% to 9% of WBCs	Phagocytic (macrophages)
Lymphocytes	Slightly larger than RBCs; nucleus almost fills the cell	25% to 33% of WBCs	Provide specific immune response and produce antibodies
Platelets (thrombocytes)	Cytoplasmic fragments; survive 5 to 9 days	1.3 to 4.0 × 10^5	Stimulates clotting; releases serotonin, causing vasoconstrictions

Table 6. Enzymes and their actions

ENZYME	CATALYZES
RNA polymerase	DNA uncoiling and synthesis of RNA
DNA polymerase	DNA uncoiling and synthesis of DNA
Protein kinase	Phosphorylation of a protein
Phosphoprotein phosphatase	Dephosphorylation of a protein
Catalase	Breakdown of hydrogen peroxide into water and oxygen
Hexokinase	Phosphorylation of glucose
ATP synthase	Synthesis of ATP through oxidative phosphorylation
Lactate dehydrogenase	Conversion of lactic acid to pyruvic acid
Glucose-6-phosphatase	Removal of phosphate from gulcose-6-phosphate
Carbonic anhydrase	Conversion of carbonic acid to water and carbon dioxide
Amylase	Breakdown of complex carbohydrates
Lipase	Breakdown of triglycerides to monoglyceride and fatty acids
Sucrase	Breakdown of sucrose into glucose and fructose

Table 7. Sequence of Skeletal Muscle Contraction

Stimulation	Motor nerve impulse	Impulse reaches axon terminal bulb in the synaptic cleft
		↓
		In voltage gated calcium channel, open calcium ions enter from synaptic cleft
		↓
		Synaptic vesicles dilate, releasing acetylcholine into synapse
Excitation		↓
		Acetylcholine bridges synapse, binding with receptors on motor end plate
		↓
		Ligand gated channels open, Na ions enter, and K ions exit, creating an electrical flux
		↓
		Voltage gated channels open, some specific for Na and others for K-action potential
		↓
		Spreads about the sarcolemma to the T tubules into sarcoplasm
		↓
		Stimulates Ca channel in sarcoplasmic reticulum to open Ca diffuse into sarcoplasm and sarcomeres
		↓
		Ca combines with troponin
		↓
		Troponin-tropomyosin shape changes active sites on actin
Contraction		Myosin head in cocked position with ADP; phosphate group retained forms cross bridge with actin
		↓
		Phosphate group released and myosin head flexes, creating "power stroke"
Recovery		ATP replaces ADP on myosin head, breaking cross bridges re-cocking myosin head
Relaxation		Impulses cease; no more acetylcholine released
		↓
		Acetylcholine dissociates from receptor; acetylcholinesterase breaks it into acetyl and choline
		↓
		ATP binds with gated channels in sarcolemma and sarcoplasmic reticulum; Ca conducted in sarcoplasmic reticulum; Na pumped out of sarcoplasm; and K pumped in sarcoplasm, re-establishing resting potential

RESOURCES

The following resources are a partial list of references used over the past twenty-plus years teaching human anatomy and physiology and used in the development of the text and tables. These resources are not inclusive, but merely the most recently used texts.

Fox, Stuart Ira. 2009. *Fundamentals of Human Physiology*. McGraw Hill.

Fox, Stuart Ira. 2010, 2012. *Human Physiology*, 12th ed. McGraw Hill.

Fox, Stuart Ira. 2013. *Human Physiology*, 13th ed. McGraw Hill.

Hole, John W., Jr. 1993. *Human Anatomy & Physiology*, 6th ed. William C. Brown.

Martini, Frederic H, William C. Ober, Claire W. Garrison, Kathleen Welch and Ralph T. Hutchings. *Fundamentals of Anatomy and Physiology*, 3rd-7th ed. McGraw Hill.

Tortora, Gerald J., and Sandra Reynolds Grabowski. *Principles of Anatomy and Physiology*. 2000. 9th ed. John Wiley &Sons, Inc.

Salidin, Kenneth S. 2006–2015 *Anatomy & Physiology: The Unity of Form and Function*, 4th-7th ed. McGraw Hill.

Silverthorn, Dee Unglaub, William C. Ober, Claire W. Garrison, and Andrew C. Silverthorn. 2007. *Human Physiology*. Prentice Hall.

Stanfield, Cindy L., and William J. German. 2008. *Principles of Human Physiology*, 3rd ed. Pearson Benjamin Cummings.

Widmaier, Eric P., Hershel Raff and Kevin T. Strang 2011. *Vander's Human Physiology: The Mechanism of Body Function*, 12th ed. McGraw Hill.

GLOSSARY

A

a used as a prefix indicating lack of, no, or without

abdomen region of the axial anatomy below the diaphragm containing the majority of the digestive system above the iliac crest

abdominopelvic region of the axial anatomy below the diaphragm containing the digestive system, parts of the urinary, and all of the reproductive system of the female

acetabulum round cavity located in the os coxae, which is the location of the articulation of the femur, translated to mean vinegar cup

acetyl prefix indicating a two-carbon organic molecule

acetylcholine a neurotransmitter in the peripheral system that can be either an inhibitory or excitatory dependent on the receptor location

acetylcholinesterase enzyme located in synapse that breaks acetylcholine into an acetyl and choline fractions, inactivating the neurotransmitter

acromial process (acromion) process located on scapula, forming the capsule with ligaments, and coracoid process, which is the acromial-coracoid articulation at the shoulder that articulates with the clavicle

actin thin contractile protein filament involved in creating movement

action potential electrical spreading along the membrane created by depolarization and repolarization in nerve and muscle tissue

adhesion abnormal bond between layers, usually encountered in the abdomen

apidocyte cells developed from fibroblasts, which store fat

adipose tissue fat-storing tissue

adrenal gland glandular tissue co-located with the kidney, producing a series of hormones and neurotransmitters

adrenocorticotropic hormone hormone produced in anterior pituitary gland, controlling the adrenal cortex

agglutination clumping of cells, usually the result of antigen-antibody reaction

agranulocytes related to leucocytes without obvious granule in the cytoplasm

albumin most abundant small protein in the blood; major component of the colloid osmotic pressure

aldosterone mineral corticoid produced in the adrenal cortex, which actively promotes the recovery of sodium in the renal tubule

alvoeli applied to lung; a small air sac where diffusion occurs with blood; small glandular tissue in mammary gland, producing milk during lactation

amine (amino) organic acid containing nitrogen as ammonia; used as building blocks of proteins

amphiarthrose articulations that are slightly movable

anatomical position body position supine with palm facing up; standing facing forward with palms facing forward

anatomy structure and relationship of body parts

anaphylactic shock depression of bodily function due to antigen- or toxin-induced reaction, creating an enhanced positive feedback, which increases the body's reaction

anterior (ventral) the front of the body in anatomical position

antidiuretic hormone produced by hypothalamus and released from posterior pituitary gland, which promotes water reabsorption from renal tubules

apneustic pattern of breathing controlled by centers in the pons

arachnoid mater the middle layer of the meninges with space above and below with a web-like appearance

arachnoid villi capillary beds in arachnoid mater in the occipital region where cerebrospinal fluid is reabsorbed

areolar tissue connective tissues with spaces and a variety of cellular material and fibers in a jelly-like matrix

arrector pili multiunit smooth muscle attached to hair follicle in the dermis

astrocyte neuroglia that support neurons by maintenance of appropriate environment to support neuron activity

atlas first cervical vertebrae articulating with occipital condyles

atom smallest unit of matter with specific characteristics

atresia related to ovaries degeneration and reabsorption of follicles; constriction of passages or absence of expected openings

atria related to the heart; the upper receiving chambers; entry way

atrial naturetic peptide hormone produced when the atria are stretched by increased volume, which inhibits aldosterone recovery of sodium, increasing urine output

atrioventricular valves related to the heart; cuspid valves located between the atria and ventricles

autonomic nervous system the part of the peripheral nervous system that control our body in response to stress without conscious action; composed of the sympathetic and parasympathetic divisions

avascular without a direct blood supply, as the epithelial tissues and cartilage

B

ball and socket articulation a freely movable articulation attached to the appendicular girdles of movement in all three planes or axis at the shoulder and hip

basement membrane an extracellular membrane formed between epithelial and connective tissue

basophil a granular leucocyte stained by the basic dye usually obscuring the nucleus

bicuspid related to the heart; the valve between the left atria and ventricle; related to teeth having two raised areas used for crushing

bilirubin the reddish pigment produced from the destroyed heme portion of the hemoglobin; stored in the gall bladder and released with bile into the small intestine

biliverdin greenish pigment first produced with the destruction of the heme portion during the recycling of hemoglobin

bradycardia slow heart rate, less than 60 beats per minute in most individuals but common in exceptionally conditioned athletes

bronchi large cartilage supported by tubular passages for air into the lungs

bronchioles smaller muscular support tubular passages for air into the lung whose diameter is controlled by smooth muscles; site of constrictions associated with asthma

buccal term used for the mouth

buffer molecules capable of absorbing or releasing hydrogen ions to maintain a limited pH range

C

calcitonin hormone produced in the thyroid gland that reduces serum calcium levels and encourages the formation of bone

calcium carbonate the mineral form of calcium found in bone that provides hardness

callus abnormal thickening associated with epidermis or in bone repair

calsequestrin protein located in the sarcoplasmic reticulum that stores calcium during the inactive period of skeletal and cardiac muscle tissues

cancellous (spongy) bone located at the end of long bones and forming interior of other bones; constructed with spaces formed by trabeculae

capillary thin, walled segment of the vascular system where diffusion and reabsorption occurs; composed of endothelial cells and supporting connective tissues with spaces and a variety of cellular material and underlying connective tissues

carbohydrate family of organic molecules that are easily broken down for energy; composed of carbon, hydrogen, and oxygen atoms in a ratio of 1:2:1

cardiac (cardio) related to or describing relationships to the heart

cardiac output the amount of blood reaching the systemic circuit within a minute

cardiovascular descriptive term used for the heart and blood vessels without the lymphatic components

cartilage type of connective tissue that has a matrix of chondroitin sulfate and is avascular with various fibers in the matrix and cells (chondrocytes) located in lacunae

cauda equina   terminal portion of the spinal cord that has become individual nerve tracts below the level of the first lumbar
vertebrae

cephalic   toward the head or the head of the body

centrum   the body of the vertebrae

cerebro   prefix used to denote the cerebrum area or component

cerebrospinal fluid   ultra-filtered plasma formed by choroid plexuses in the third and fourth ventricles of the brain

ceruminous gland   a modified sweat gland located in the external auditory canal that produces cerumen (ear wax)

chemoreceptors   types of neural receptors that respond to chemical changes involved in taste and smell as well as other sensory functions internally

cholecystokinin   local hormone produced by the small intestine in response to fats stimulating the gall bladder; suppresses the hypothalamus

choroid   vascular layer of the eye

choroid plexus   capillary beds associated with ependymal cells, which produce cerebrospinal fluid

compact bone   bone with dense mineral crystals supported by collagen fibers in which the osteocytes are organized into osteonic systems for support

chondroitin sulfate   extracellular matrix found in the cartilage

collagen   fibrous protein found in connective tissues providing strength

colostrum   antibody-enriched production from the mammary glands for about three days after the birth of a child; provides temporary immune protection

condyle   articulation surface found in freely movable joints (synovial joints)

condyoid (ellipsoidal) joint   freely movable articulation between an oval-shaped condyle that articulates with an elliptical cavity biaxial movement

conjunctiva   delicate mucus membrane surrounding the eye, isolating the posterior orbit

contractile   ability to shorten, creating tension as in muscle tissue

conus medullaris   tapered portion of the spinal cord between the twelfth thoracic and second lumbar vertebrae, giving rise to nerve tracts (cauda equina)

coracoid process   anterior process from the scapula where ligaments attach and form the acromial coracoid articulation capsule

coronal (frontal)   a plane dividing the body into an anterior (ventral) and a posterior (dorsal) portion along the long axis

corpus   a body

corpus albicans   the remaining location of a follicle in an ovary filled with collagen

corpus callosum   mass of myelinated axons connecting the cerebral hemispheres

corpus luteum   the yellow body of the remainder of the follicle following ovulation, which continues to be functionally capable of producing hormones

cortex   outer layer of an organ

costa (costal)   referring to ribs

covalent bond type of bonds between atoms in which the electrons are shared; found in organic molecules

cranium (cranial) bony encapsulation of the head that contains the brain

cutaneous the outer membrane of the body; also called skin or integument

cyto pertaining to cells

cytokines small proteins that act as chemical messengers of short range from where they are produced

cytology studying cells

cytosol contents of a cell including the organelles and cytoplasm

D

deamination the process of removal of an amine group (NH2) from an amino acid in order for an amino acid be used as an energy

dendrite extension of a neuron that receives impulses that are conducted to the cell body

depolarization a change in the electrical potential, creating an action based on the shifts of the concentration of ions across a membrane found in neurons and muscle cells

derma (dermal) pertaining to the skin

dermatologist medical professional who specializes in the disease and disorders of the skin

dermis the layer of the skin; vascular and composed of connective tissue

diapedesis the passing of macrophages between the endothelial cells of the capillary; for entry into the interstitial spaces in tissue

diaphragm the skeletal muscle involved in breathing, separating the thoracic spine and the abdomen

diaphysis the end of long bones composed of spongy bone

diarthrose articulations that are freely movable and lubricated by synovial fluid; has a synovial capsule surrounding the articulations

diastole term applied to the heart when the cardiac chambers are in a refractory interval

differentiation the separation of cells from the embryological cells that then become specialized in function

diffusion passive movement of molecules from a higher to lower concentration based on molecular dynamics with an environment

distal applied to skeletal components that are further from a reference point; usually related to the axial skeleton

dorsal (posterior) the back above the vertebral column

dorsal cavity major cavity system in the body; divided into a cranial and spinal (vertebral) divisions

duodenum the first segment of the small intestine receiving partially digested food from the stomach acid being neutralized and enzyme-aided to promote further digestion; approximate length is 10 inches

dura mater the tough outer meniges that surrounds the brain

dust cells macrophages found in the alveoli that remove debris and protect the alveoli

E

edema   accumulation of fluids in tissue; swelling due to injury

elastic   ability to return to shape following stretching or compressing

elastin   a fibrous protein that is stretchable and returns to shape

end diastolic volume   the volume of blood in the left ventricle before contraction

endo   within

endocardium   the interior layer of the heart consisting of the endothelial lining and supporting connective tissue

endochondral   a type of ossification that begins as cartilage around which bone forms

endocytosis   formation of internal vesicles from the plasma membrane

endometrium   the inner layer of the uterus composed of a basal layer that replaces the functional layer, which is shed with menses during each cycle

endomysium   thin layer of connective tissue surrounding skeletal myofibrils isolating each myofibril from each other and conducting the generated tension

endoplasmic reticulum   a bilayer organelle found in cells where lips and lipoproteins are produced and ions are stored in muscle tissue

endothelial   epithelial tissues found as lining of structures in the body, specifically blood vessels and heart with squamous cells; columnar in the airways

end systolic volume   the volume of blood retained in the left ventricle following ejection into the systemic circuit

enteric   a collective term used for the small intestines

ependymal   glial cells that are responsible for the production and circulation of cerebrospinal fluid

epidermis   the outer portion of integument and cutaneous membrane composed of stratified and keratinized squamous epithelial cells

epigastric   term used for division of the abdominal region into nine regions; the region superior to the central umbilical region below the diaphragm

epimysium   outer layer connective tissue that covers a skeletal muscle and merges with connective tissues to form the tendons that attach muscle to bone

epinephrine   a neurotransmitter found in the sympathetic nervous system that acts as excitatory or inhibitory in a variety of locations

epiphyses   end of long bones composed of spongy bone that house the blood forming stem cells

epiphyseal plate   located in the transition zone of the long bone where length growth occurs until approximately the mid 20's

epithelial   type of tissue with close gap junctions that cover line structures

erythrocytes   the red blood cells that contain hemoglobin transporting oxygen; mature in the bone marrow losing both the nucleus and mitochondria

erythroblastosis fetalis   immune condition when Rh positive antibodies have been produced in a mother who is Rh negative that results in fetal death

erythropoietin   hormone produced in the kidney that promotes increased production of erythrocytes in the bone marrow; two- to three-day response to change in concentration

estrogen   hormone produced by the ovaries which stimulates secondary female
structures and regulates reproduction; made up of six different variants, three of which are beta estradiol, esterone, and estriol

expiration (expiratory volume)   the removal of air from the airway; produced by the relaxation of muscles of the diaphragm as well as those of the rib cage

eukaryote   cells that possess a membrane bound

F

facilitated diffusion   diffusion of non-lipid soluble substance across the membrane-aided carrier (transporter) protein in the selectively permeable
membrane

fascia   the fibrous membrane surrounding, supporting, and separating muscles aided by a carrier protein

fascicle   bundle of myofibrils or nerve tracts

fenestrated   having window-like openings in capillaries

ferric   form of iron ion having a charge of plus three

ferrous   form of iron ion having a charge of plus two, which is the form used to produce hemoglobin heme fraction

fetus   term referring to a developing embryo after the first trimester

fibrinogen   soluble form of protein in plasma that converts to insoluble fibrin in

clotting, which attaches to damaged tissue creating the clot

fibro cartilage   cartilage containing large bundles of collagen fibers in the matrix

follicle-stimulating hormone   hormone produced in the anterior pituitary that stimulates the development of ovarian follicles in females and stimulates Sertoli cells to produce androgen-binding protein

fovea centralis   center of focus in the eye surrounded by the macula lutea that is dominated by cones and has the sharpest color vision; revives light directly unlike most rods and cones

fracture   referring to a break in a bone

frenulum   mucus membranes found in the oral cavity that connect two parts, limiting movement

frontal (coronal) suture   large sutural articulation between the frontal and parietal bones of the cranium

G

galanin   neurotransmitter that stimulates cravings for fats

gastric   referring to the mucosa, the stomach

gastrin   hormone stimulating the production of gastric juices and stomach motility

gastro ferritin   a protein produced in the stomach that binds with iron in the ferrous form, permitting absorption in the small intestine

gladiolus   the body of the sternum where the majority of the true ribs articulate

glia   shortened name for neuroglia that provide support to neurons

globulin   globular protein that can be an enzyme or antibody; proteins found in blood plasma fraction

glomerulus   spherical mass of capillaries in the kidney; spherical mass of nerve ends located in the olfactory bulb

gluconeogenesis   process of forming glucose from sources such as fats or amino acids

glutamic acid (glutamate)   amino acid neurotransmitter involved with learning and memory

glyco   pertaining to carbohydrate

glycogen   energy storage polymer of glucose found in liver and skeletal muscle

glycolysis   anaerobic decomposition of glucose that produces limited ATP and two molecules of pyruvic acid (pyruvate)

glycoprotein   protein with a small carbohydrate attached covalently

Golgi complex   complex that modifies and packages proteins and produces carbohydrates

gomphoses   immovable articulation found with a socket and cone bound by a fiber as in the teeth

granulocyte   grouping of leucocytes that have large granules in the cytoplasm which stain with either the acid or basic dyes

granulosa cells   cells that line the ovarian follicle

gravid   term used for pregnancy

H

heme   non-protein fraction of hemoglobin that contains iron

hemopoiesis   general term for the production of blood-formed elements

hilum   midpoint of a concave structure where blood and lymphatic vessels enter and exit along with nerves

histo   pertains to tissue

histology   the study of tissue

homeostasis   maintaining a stable internal environment

human chorionic gonadotropin   hormone produced by the chorion that stimulates the corpus luteum functioning and that produces hormones to sustain pregnancy

humoral   refers to the immune system subdivision that produces and protects the body by the B lymphocytes and that produces plasma cells that produce antibodies

hyaline   referring to cartilage with a clear matrix that contains very fine collagen fibers in a jelly-like matrix

hyaluronic acid   found in connective tissue where it forms a gel-like matrix in tissue

hypertonic   environment having an osmotic pressure higher than the cell

hypotoinc   environment with an osmotic pressure lower than the cell

hypogastric   region of the abdomen directly below the umbilical region in nine-region abdominal system

hydroxyapatite   calcium phosphate molecule found in bone

hypoxia   reduced partial pressure of oxygen

I

ileocecal valve   valve separating the small and large intestine that terminates the ileum and empties into the cecum

ileum   forms about 60 percent of the small intestine, which is involved with the completion of digestion and the large absorption of nutrients

illium   largest region of the os coxa

immunity   ability to defend against specific infective agents

implantation   the process of attachment of the conceptus to the uterus

inferior   below some reference point

inflammation   response of tissue to injury or infection indicated by redness, heat, swelling, and pain

inhibin   hormone produced by corpus luteum that regulates follicle-stimulating hormone release

insula   fifth lobe of cerebrum in the lateral sulcus involved in taste, pain, emotion, empathy, heart rate, and blood pressure

integument   alternate name for the skin or cutaneous membrane with its accessory structures

intercalated dis   tight gap junction in cardiac muscle that promotes cell-to-cell stimulation

interferons   proteins secreted by infected cells prior to death to alert the immune system

internal   refers to an innermost layer or region

interstitial   extracellular spaces within tissues; usually fluid filled

intramembranous ossification   formation of bone in fibrous connective tissue, such as the flat bones of the skull

inspiration   moving air into the airway, ventilating the system

ion pump   term applied to integral proteins that are capable of using energy of ATP to move ions against a concentration gradient

iris   the color portion of the eye that possesses muscle to control the pupil for control of light entering the lens and retina

irritable (excitable)   term used for muscle tissue that indicates it is capable of stimulation and response by experiencing an action potential

ischium   portion of the os coxa located inferior to the ilium and posterior to the pubis regions

J

juxta   beside or adjacent to

K

keratin   tough, waxy protein that forms a resistant barrier to microbial attack

keratinization   process of formation of keratin and filling epithelial cells with keratin

keratinocytes   epithelial cells capable of producing keratin and becoming the stratum corneum of the epidermis, hair shafts, and nails (claws)

L

lacrimal glands   glandular structure associated with the eye, producing tears

lamina   thin plate or layer such as the area of bone over the vertebral foramen of the vertebrae

leukotrienes   mediators of allergy and inflammatory responses

lipase enzyme that assists in the breakdown of lipids

M

macula lutea area surrounding the fovea of the eye that is composed mostly of cones

melanocytes cells associated with the stratum basale that produce melanin which is transferred to keratinocytes

melanin brownish to black granular pigment capable of reducing ultraviolet passage in the skin

melatonin Produced by pineal gland and regulates the 24-hour cycle

N

nociceptor general term related to pain receptors that detect tissue damage

O

organ of Corti structure found in the scala media of the cochlear that consists of a basilar membrane, inner and outer hair cells, and tectorial membrane that responds to pressure waves in in the scala tympani

osteoblasts bone cells capable of formation of new mineral matrix and collagen

osteoclasts bone cells capable of breaking down bone by removal of mineral matrix and collagen

osteocytes undifferentiated bone cells

osteonic system system located in compact bone that is arranged around a central canal that provides nutrients and waste removal from osteocytes

oxytocin hormone released from posterior pituitary gland that stimulates smooth muscle contraction intensity

P

pampiniform plexus A combination of arterial vessels and venous vessels where a heat exchange occurs to protect heat sensitive tissues; for example the testes in humans

papilla little hill or convolutions that increase the surface area between two layers (epidermis and dermis) for exchange of nutrients and waste

parenchyma the physiologically active component of an organ or tissue

parietal name of flat bones of skull or the layer of serous tissue on the surface of an organ

pericardial double-walled membrane surrounding the heart

peritoneum serous membrane lining the abdominal cavity

pleura double-walled membrane surrounding the lungs

pneumothorax term referring to a penetration of the pleura, which collapses the lung

prokaryote group of microorganisms that lack a distinct membrane-bound nucleus

prolapse term referring to incompetent valve, allowing blood to flow reverse

pyrexia elevated temperature or fever

Q

quadri, quadro prefix referring to four

R

renin enzyme produced in the kidney that converts plasma protein into angiotensin, raising blood pressure

reticulocyte form of erythrocyte that retains chromosomal material in peripheral circulation, indicating rapid movement of erythrocytes from bone marrow

rugae ridged or folded lining found in the stomach that promotes mixing in the relaxed urinary bladder

S

sarco prefix referring to meat or muscle

scrotum male structure housing testes exterior to abdominal cavity

sebaceous glandular structure that produces sebrum (oil) composed of fatty acids that creates an acid envelop on the skin

septic a form of shock produced from toxins produced by microbes

sertoli cells (sustentacular) provide support to the germ tissue of testes and produce androgen-binding protein, regulating sperm production

serous (serosa) double-layer membrane lining of closed body cavities that produces serous fluid

serum (sera) plasma that has been allowed to convert the fibrinogen into fibrin, (clot) removing the formed elements and clotting components

soma term applied to the central part of cell housing the nucleus or the central portion of the vertebra meaning body

squamous referring to a flattened cell shape or suture in the skull

stenosis refers to a stiffened condition of the myocardium or the valves, requiring additional work to accomplish activity

synovial the capsule surrounding freely movable articulation and containing synovial fluid for lubrication

systole (systolic) term for the contraction of the myocardium

T

tachycardia heart rates above 100 beats per minute severity; related to how much above that level is important

transferrin plasma protein that binds with iron two-plus moving it to liver and bone marrow

tropomyosin regulatory protein on muscle fibers that cover the myosin binding site on the actin strand

troponin regulatory protein on skeletal and cardiac muscle that combines with calcium and permits tropomyosin to rotate exposing the myosin binding site on actin

U

uterine refers to some action of the uterus

uvula projection from the soft palate, keeping food in mouth until swallowing; also reported as center of gag reflex

V

vas referring to vessel or duct

ventral (anterior) referring to the front of the body from quadrupeds

viscera (visceral) organs of the internal body spaces

vulva general inclusive term for female external genitalia

Y

yolk sac embryonic membrane giving rise to first blood

Z

zygote single fertilized cell or conceptus

www.ingramcontent.com/pod-product-compliance
Lightning Source LLC
Chambersburg PA
CBHW082034230326

41598CB00081B/6509